# DETAILED REPORT

OF THE

## PROCEEDINGS HAD IN COMMEMORATION

OF THE

## SUCCESSFUL LAYING

OF THE

# ATLANTIC TELEGRAPH CABLE,

BY ORDER OF THE

## Common Council of the City of New York.

---

BY C. T. McCLENAHAN.

---

**NEW YORK, 1859.**

NEW YORK:
EDMUND JONES & CO., CORPORATION PRINTERS,
No. 26 John Street.
1863.

*Whereas*, The report of the Special Committee appointed for the purpose of making suitable arrangements for the celebration incident to the laying of the Atlantic Telegraph Cable never was presented, except in manuscript form; and, inasmuch as the same is of great importance to the public, as well as to the scientific world, and should be printed for the information of the public; therefore, be it

*Resolved*, That five thousand copies of the report of said Special Committee be printed and bound, by the printer to this Board, under the direction of the Clerk, for the use and benefit of the members of the Common Council and the public.

Adopted by the Board of Councilmen March 22, 1860.

Adopted by the Board of Aldermen May 7, 1860.

Received, June 8, 1860, from his Honor the Mayor, without his approval or objections thereto; therefore, under the provisions of the Amended Charter of 1856, the same became adopted.

# PREFACE.

As the event whose successful issue we have just celebrated is one that must ever rank among the most important of the Nineteenth Century, whatever may be its present apparent and intrinsic value in so far as the actual enterprise *in itself* may be considered, the Joint Special Committees of the Common Council have deemed it not inappropriate that the commemoration of this instance of perseverance, energy, and skill should become a matter of history.

By it the practicability of submerging a Submarine Cable of any required length, and the fact, that intelligence can be transmitted through such Cable, have been demonstrated beyond a cavil, leaving it only for time to illustrate and produce, through the inventive genius of this most enterprising age, the nature and kind of material which shall be the best and most efficiently adapted to resist the opposing and corroding elements of nature, to crown any future efforts with the most complete success in every respect.

The Committees have, therefore, directed a careful account of all that might be interesting in connection with the celebration to be prepared, and, with the consent and under the authority of the Common Council, published in detail.

The question as to the benefit to be derived from the Cable now laid, *per se*, is one that it is not proposed to discuss, nor was the celebration intended to be a manifestation of rejoicing at the success of a private enterprise; but the design was to show forth the feeling and appreciation of the public of a great triumph of art and science over the obstacles and difficulties placed by nature in the path of the onward progress of human knowledge and skill.

The conception of the enterprise was never equaled in the annals of daring undertakingsn. In no records do we find such stupendous obstacles attacked; and nowhere in history do we discern such unflinching perseverance and uncompromising energy displayed.

The result has been the clear demonstration, that the feat of crossing the seas and large oceans by means of telegraphic Cables *can* be accomplished, and already the impetus that has been given to the extension of telegraphic communications in all parts of the world—thus annihilating space and drawing the nations together in closer bonds of brotherhood—promises fair to hasten the coming of that time when all men shall dwell together in unity, and wars and rumors of wars shall cease to be.

NEW YORK, September, 1858.

On the 5th of August, it was announced, by a telegraphic dispatch from Newfoundland, that the great Atlantic Telegraph Cable had been finally and successfully landed at Trinity Bay, and that signals had been transmitted hence, through the Cable, to Valentia Bay, on the coast of Ireland.

In consequence of the foregoing advices, the following call for a meeting of the Board of Aldermen was issued:

New York, August 7, 1858.

Sir—The President being absent, it has been suggested by several members of the Board of Aldermen to make certain arrangements in commemoration of the successful aying of the Telegraph Cable. You are, therefore, requested to attend an informal meeting on Monday, the 9th instant, at room No. 12, City Hall, at 4 P. M.

By order,

D. T. VALENTINE,
*Clerk.*

And, in pursuance of this notice, the Board convened on the 9th of August, when the following message from his Honor the Mayor was received and read:

Mayor's Office,
New York, August 9, 1858.

*To the Honorable the Common Council:*

Gentlemen—The great work of uniting Europe with America—the New with the Old World—by means of the

Electric Telegraph, has been successfully completed ; and it becomes, in my opinion, the authorities of this city to adopt suitable measures for a public celebration of so important an event.

I have, therefore, deemed it my duty to communicate with your Honorable Body, as well for the purpose of making such recommendation, as to congratulate you, and, through you, the citizens of New York, on the complete triumph which superior energy and perseverance have accomplished in uniting together, by the Atlantic Cable, not only our own city, but the whole of our country, with Europe and the greater part of the civilized world.

The important and beneficial results to our race which this great event promises cannot be wholly anticipated ; but that it will tend to the perpetual peace and increased happiness of the two leading nations who have joined in the labor and cost of the enterprise, cannot be doubted, while itself the offspring of science and that civilization which is founded on Christian principles, it announces to the whole world the reign of lasting peace and good-will to all men.

Our city and country have a right to claim an ample share in the glory of this peaceful achievement. The genius of Franklin, the patriot and philosopher, lit the way to the brilliant succession of discoveries in electrical science, the most useful of which is undoubtedly the practical application of this noble element to telegraphic purposes by our countryman, Morse—thus enabling the actions, feelings and sentiments of every people to be communicated to each other with almost the rapidity of thought. Besides, the majority of the active officers of the Company, by whose public spirit and noble enterprise the electric chain has been laid, are our own citizens and countrymen. Their names are known, and, if we do not, posterity will surely reward them. While the noble ship, the Niagara, was generously tendered by our Government to aid in the important work, by the consummate skill and sleepless care of her commander and other officers, and the eminent and skillful gentlemen on board of her, the western part of the Cable was laid.

On an occasion like this, in the celebration of such an event, I am sure you, on behalf of our city, will act in no envious spirit, and that all who have contributed to this result will be fitly remembered, and that you will do ample justice to both the living, who accomplished it, and the genius of the past, which animated these exertions.

I most respectfuly recommend, in conclusion, that, among other arrangements, the hospitalities of the city be extended to the officers of the Niagara, and of the national vessels of Great Britain connected with her in laying the Cable as well as to the eminent gentlemen whose skill and energy contributed to the glorious accomplishment of the work.

I would also recommend that our citizens be requested to illuminate their houses, and that the Common Council çause the public buildings to be illuminated on the evening of the day you may fix for the public celebration.

DANIEL F. TIEMANN,
*Mayor.*

The following preamble and resolutions were then offered by Alderman McSpedon, and unanimously adopted:

*Whereas*, The truly gratifying intelligence having been

communicated to our people of the successful laying of the Atlantic Cable, and the final triumph of this, the greatest and boldest project of the age; and,

*Whereas*, This grand and proud event should be everywhere publicly acknowledged, and the projectors of this gigantic enterprise appropriately commended for their Herculean labors, which so happily have ended in a complete success and triumph; therefore, be it

*Resolved*, That the thanks of the Common Council are eminently due, and are hereby gratefully tendered to our distinguished and universally esteemed fellow-citizen, Cyrus W. Field, Esq., and also to his associates of the New York, Newfoundland and London Telegraph Company, and their numerous and valuable assistants, to whom we are indebted for the accomplishment of this wonderful and prodigious enterprise, connecting the Old and New Worlds, which was deemed by many impracticable and impossible, and which must lead to great and important results between the two nations that are now incalculable, and which, it is believed, will prove a mutual advantage to the civilized world.

*Resolved*, That, in commemoration of this world-renowned achievement, the Common Council tender to the officers of the Telegraph Company, and such other gentlemen as were engaged in this inconceivably great enterprise, a municipal dinner, at such time as may be to them convenient and acceptable.

*Resolved*, That, in further commemoration of this glorious success, the City Hall be illuminated, and that suitable fireworks for the occasion be procured to give additional expression to the general rejoicing of our entire community.

*Resolved*, That, as we owe the result of this heroic event principally to our own fellow-citizen, Cyrus W. Field, Esq., (whose master-mind, energy and perseverance, amid doubt and disaster, finally triumphed), that he be requested to sit for his portrait, to be placed in the Governor's room, in the City Hall.

*Resolved*, That a Committee be appointed to carry out the foregoing resolutions.

*Resolved*, That the Clerk of the Common Council cause copies of the above proceedings to be appropriately engrossed and furnished to each of the officers connected with the expedition.

Aldermen Boole moved that the Special Committee to be appointed in accordance with the foregoing resolutions should consist of five, which was carried, and

Aldermen McSpedon, Hoffmire, Tucker, Boole and Lynes, were appointed to serve as such Committee. Subsequently, Alderman Tucker resigning, Alderman Owens was appointed in his stead.

The action of the Board of Aldermen was concurred in by the Board of Councilmen at their regular meeting on the evening of the 10th of August, as will appear from the following extract from the Journal of the latter:

"From the Board of Aldermen:

"Communication from his Honor the Mayor, with preamble and resolutions expressive of thanks to Cyrus W. Field, Esq., and his associates in laying the Atlantic Telegraph Cable, and tendering to him and the officers of the Telegraph Company a municipal dinner, at such time as may be convenient to them; also, appointing a Special Committee to carry out the resolutions.

"Concurred in, and Councilmen Dunn, Mulligan, Bunce, Ross, and Bickford were appointed such Committee."

On the 20th of December, 1858, the Joint Special Committee made the following report:

The Special Committees appointed by the Common Council to arrange and supervise the celebration in honor of the successful laying of the Atlantic Telegraph Cable, by appropriate ceremonies, and a municipal dinner in honor of Cyrus W. Field and others who had been engaged in the undertaking, in accordance with resolutions of 9th and 10th of August, would respectfully

REPORT:

That, with a full appreciation of the magnitude of the great work, whose recent successful termination it was the design of the Common Council to celebrate, the Committees entered with energy and great interest upon the duties assigned to them.

They fully realized that it was no ordinary occasion they were called upon to honor; it was one calling forth the admiration, the wonder and the enthusiasm of the whole world; one in which much of the future welfare of the entire human race was concerned; an achievement so vast that the most sanguine had been surprised by its success; for by indomitable will, energy and perseverance the electric fluid, whose mighty force, until the time of the great Franklin, had been held in awe and looked upon in terror, and which had been chained and tamed to subserve the purposes of man by Morse, was now made by Field to bridge the broad Atlantic, and by its silent influence to annihilate the distance that separates us from our brethren beyond the seas.

The miracle of this age of miracles had been wrought! What mind can conceive the mighty effects to flow from this practical demonstration of the power of man over the subtle element? Who can foretell the changes it is destined to work in the polity of the world? And, will it not mark the commencement of an era of the most serious revolutions —revolutions that through its means will, in all probability, be as bloodless as they will be important and startling by their rapidity and extent!

These are questions that have occupied the public mind since the announcement that the Atlantic Telegraph Cable had been finally and safely landed at Trinity Bay, and the universal interest in the matter has been fully attested by the intense enthusiasm that has pervaded all classes throughout the whole United States and the British possessions.

The most immediate, and, for the present, important influence of this great triumph of mind and genius over the most formidable obstacles of nature, so far as our own city is concerned, is necessarily that which it will have upon commerce; and upon the trade of New York depends, in a material measure, the prosperity and advancement of that of the Union. In all commercial matters New York is regarded as the exponent of our people; it is ever expected to take the initiative, and its enterprise is relied upon in the execution of great achievements. The pride of New York is the pride of the nation, and it therefore behooved the Committees, in carrying out the instructions and designs of the Common Council, to do so in a manner that should prove, not only satisfactory to your Honorable Body, but also in a manner that should be worthy of this great metropolis—to whose citizens and their enterprise the conception, the attempt, and the successful issue of this, the greatest undertaking of the nineteenth century, are to so great an extent due—and make the celebration one that should mark the estimation in which the accomplishment of this great event is held by this community.

Upon the fact being fully established that the Old World had in reality been joined in union to the New by the bonds of the great Atlantic Cable, the Committees deemed it incumbent upon them, in following their instructions, to celebrate by public rejoicings the landing of the Cable, reserving the general demonstration until such time as Mr. Field, and the others who had taken part in the great work, could participate therein. They, therefore, decided that upon the transmission of the President's reply to a message that it was understood would be forwarded to him by the Queen of Great Britain, a grand salvo of one hundred guns should be fired at the Park, and national salutes at the Battery and at Central Park; also, that on the evening of the same day the City Hall and other public buildings should be illuminated, and that tar-barrels should be burned at the Battery and at various points along the North and East Rivers.

They also provided for a handsome and appropriate display of fireworks to take place in front of the City Hall, accompanied with music by Dodworth's brass band.

The occupants of buildings in the vicinity of the Park were also requested to illuminate them on the same evening, which request was very handsomely complied with.

The salutes were conducted by the Scott Life Guard, under orders from General J. H. Hobart Ward, on the 17th of August, at noon, in handsome style, and without accident. The Guard also fired a salvo of one hundred guns at sunrise of the same day.

The pyrotechnic display was furnished by J. Edge, Jr., and, together with the music, afforded very general satisfaction to the public.

It is a matter of very serious regret, however, that, notwithstanding the extraordinary precautions against accident taken by the Committees, fire was, by some means, communicated to the cupola of the City Hall, and the upper part of that noble edifice destroyed.

It being understood that there would be a grand celebration throughout Great Britain on the occasion of the laying of the Atlantic Cable, it was considered desirable that the festivities on this side of the Atlantic should take place upon the same day, if possible ; but not being aware of the precise time fixed upon on the other side of the ocean, the Committees, after consultation with Mr. Field and his Honor the Mayor, concluded to concur in the opinion of the former, that the 1st of September would be the most suitable day for the grand celebration, and that the dinner should take place on the day following.

Mr. Field's further suggestion, that a dispatch should be sent to the Lord Mayor of London, informing him of the selection of that day, so that the cities of Europe might time their rejoicings to take place simultaneously with our own, was also adopted, and notice to that effect was forwarded to the Lord Mayor of London, by a telegraphic dispatch from his Honor Mayor Tiemann.

After several meetings and due consideration, the Committee resolved upon the following order of proceedings to be observed in honoring the occasion :

A reception by the Mayor and civic authorities, at the Battery, of Mr. C. W. Field, Captain Hudson and other officers who had been engaged in the laying of the Cable, during which a salvo of one hundred guns should be fired at the Battery, and salutes from the City Hall and Central Park; to be followed by a grand procession, consisting of the guests, United States military and naval officers on the station, the military and fire departments, civil authorities, societies, trades, &c., to proceed to the Crystal Palace, where appropriate addresses and testimonials should be presented to Cyrus W. Field, Captains Hudson, of the Niagara, Preedy, of the Agamemnon, and Dayman, of the Gorgon, and to Engineer W. E. Everett, of the U. S. Navy, together with addresses to Captains Adlum, of the Valorous, and Otter, of the Porcupine, and Engineer Woodhouse, of the English Navy (all of whom had been engaged in the great work), and to the New York, Newfoundland and London, and the Atlantic Telegraph Companies.

Mrs. Ann S. Stephens having kindly prepared a couple of odes suitable to the occasion, and the Harmonic Society having generously proffered the services of their extensive and excellent choir to perform the same at the Crystal Palace, it was proposed that the ceremonies should be varied not only by instrumental, but also by vocal music, into which should be introduced the odes of Mrs. Stephens, together with appropriate extracts from oratorios, &c., and the offer of the Harmonic Society was accepted for the performance of this part.

The festivities to conclude with a grand torchlight procession, to be composed of the firemen of New York and such societies as chose to assist therein, to march from the Palace to the City Hall, where a brilliant and unsurpassed display of fireworks was generously proposed by Mr. G. A. Lilliendahl, to be prepared at his own expense, as a fitting finale to the rejoicings of the day.

The residents and others along the line of the route of torchlight procession were also requested to illuminate their buildings, and thus add to the general effect.

In pursuance of the instructions to the Committee, it was determined that the municipal dinner to be given to Mr. Field, Captain Hudson and the others who had participated in the arduous undertaking of submerging the cable, should take place on the 2d of September, at the Metropolitan Hotel, to be got up in appropriate style, and that such guests should be invited as by their presence would assist in rendering the fête one worthy of the great occasion intended to be commemorated.

Mr. J. W. Hadfield having also generously offered to furnish, gratuitously, a handsome pyrotechnic display, to be exhibited at the Park, his offer was accepted and the exhibition set for the same evening.

All these dispositions were made known to the parties more immediately interested, and their approval and readiness to co-operate in the same signified to the Committees.

Wishing to make the celebration as universal as possible, the different cities, towns and villages throughout the United States and British possessions, were notified that the day therefor was fixed for the 1st of September, and they were requested to have their respective festivities take place simultaneously, which request, so far as your Committee have as yet learned, was complied with wherever any ovation was made.

The President and Vice-President of the United States and Cabinet, Governors of States, officers of our own State and city Government, U. S. naval and military officers on this station, Captain Dayman and officers of H. B. M. steamer Gorgon, Rear Admiral Sir Houston Stewart and the principal officers of his flag-ship Indus, foreign Ministers at the Federal Capitol and Consuls at this city, Mayors and Bishops of the various cities of Canada, Newfoundland and New Brunswick, together with such members and ex-members of the Federal legislative body as had contributed in any measure towards the success of the enterprise, and many other distinguished individuals, were invited to take part in the festivities of the 1st of September and to participate of the dinner of the day following.

The Committee regretted that his duties prevented Rear Admiral Sir H. Stewart from visiting New York at that time; but he was ably represented by Flag-Lieut. Aug. Kingston, who, together with four of the officers of the Indus, Captain Dayman and the officers of the steamer Gorgon, were received by the Committee upon their arrival from Halifax, with all due attention, and tendered the freedom and hospitalities of the city.

Some of the most gifted of America's talented sons were requested to prepare addresses to be delivered at the Crystal Palace to the gentlemen designated, and the request was most kindly and cordially complied with, while others who were called upon to take part in the other ceremonies at the Palace and at the dinner, acquiesced with great alacrity and cheerfulness.

A special messenger was dispatched to Washington to deliver the invitations to the high Federal authorities as an attention due to their position.

In view of the great benefits to be conferred upon the cause of Christianity as well as upon civilization, throughout the world, by means of this great application of the discovery of Morse, the trustees of Trinity Church volunteered to join in the festivities of the 1st September by having the beautiful choir and splendid organ of that church perform a grand "Te Deum," to which the authorities were invited, under the direction of the Rev. F. Ogilby, Rector, the edifice being at the time handsomely and appropriately decorated.

The Committee feel great pleasure in reporting to your Honorable Body the extremely satisfactory manner in which the entire programme was carried out, not a single accident occurring to mar the enjoyment of the festal arrangements, and they flatter themselves that in the execution of the designs and instructions of the Common Council, they have, as was the intention, maintained the character of New York for generosity in recognizing and rewarding merit, genius and enterprise. As the great representative of our Union, she could do no less than make the celebration we have had the honor to superintend, in a manner commensurate with the vastness of the scheme whose successful accomplishment it was intended to commemorate, and in which she is so intimately and immensely interested

To Brigadier-General Charles Yates, who kindly consented to act as Grand Marshal of the day, is mainly due the credit for the orderly and pleasant manner in which the procession passed off, its excellent arrangement having been entirely planned by him; and much praise must be accorded to the Chief Engineer, Henry H. Howard, of the Fire Department, who, as marshal of the evening, arranged and superintended the admirable torchlight procession, which has never been surpassed in this city, and which formed the escort from the Crystal Palace to the City Hall on the night of the 1st of September. To both of these gentlemen the Committees feel themselves greatly indebted for their valuable assistance, and would beg to express their acknowledgments therefor.

And now, before concluding this report of their proceedings, may your Committee express their hope that this wonderful achievement of the progressiveness of the nineteenth century may prove, not only the prototype of eternal union between the nations that it links together, the harbinger of universal peace and destroyer of war, but also that, under the blessing of that Divine Providence which has so signally favored our generation in permitting its accomplishment in this, our day, it shall be the means of shedding light to the benighted, of bestowing the blessings of Christianity upon the heathen, and of bringing the whole world together into those bonds of love and brotherhood conducing unto the "perfect end," and so shall the names of Franklin, Morse, and Field ever descend to future generations, to be breathed together with reverence as preeminent among the benefactors of the human race.

In conclusion, your Committee would state that they have exercised every economy consistent with the scale upon which it was desirable that the celebration should be conducted, and the expenses have been considerably decreased by the generosity of sundry public-spirited individuals, to whom due acknowledgments should be made. Your Committee would, in this connection, mention

Gust. A. Lilliendahl, and J. W. Hadfield, for their gratuitous and splendid displays of fireworks at the City Hall; the former on the evening of Wednesday, September 1, and the latter on the evening of the succeeding day;

W. Hall and others, lessees of the Crystal Palace, for their kindness in furnishing the use of the Palace for the Celebration on the 1st September, free of charge;

The Harmonic Society, for their services in "discoursing sweet music" at the Crystal Palace, on the 1st September;

Commodore L. Kearney, Commandant of the New York Navy Yard, for his kind tender of flags, and for the use of a band of music furnished for the procession;

Superintendent Talmadge and Commissioners of Police, for their courtesy in detailing police forces, at the request of the Committees ;

Adjutant-General F. Townsend and Commissary-General Ward, for their promptitude in furnishing ammunition and detailments for the salvos and salutes; and

Bouché Fils & Drouet, for a present of fifteen baskets of champagne, to be used at the dinner at the Metropolitan Hotel ;

The different trades and societies, for their assistance generally in giving eclat to the display on the first of September.

To R. W. Lowber, Esq., the Committees would tender their most sincere thanks and warmest acknowledgments for his very kind, efficient and valuable aid throughout the whole period of their active duties.

Notwithstanding the generous contributions mentioned in the foregoing, the expenses have necessarily been large, and the Committee estimate that the sum of twenty-five thousand dollars will be requisite to satisfy the same, the which sum they would request should be appropriated therefor by the adoption of the annexed resolution.

The Committee would respectfully add that an extended report of the celebrations, with an account of all matters of interest connected with the Atlantic Cable, will be prepared in book form in the course of a short time, and presented for the approval of the Common Council.

*Resolved*, That the sum of twenty-five thousand dollars

be, and the same is hereby, appropriated to defray the expenses incurred by the recent celebration of the laying of the Atlantic Telegraph Cable, and that the Comptroller be, and he hereby is, authorized and directed to pay such bills therefor as may be approved by the Committee on Celebration, and the Street Commissioner, amounting in the aggregate to not over that sum.

| | |
|---|---|
| THOS. McSPEDON,<br>HENRY HOFFMIRE,<br>F. I. A. BOOLE,<br>JOHN LYNES,<br>JAMES OWENS, | *Committee on Celebration, of Board of Aldermen.* |
| T. A. DUNN,<br>S. A. BUNCE,<br>GEO. P. BICKFORD,<br>GEORGE ROSS, | *Committee on Celebration, of Board of Councilmen.* |

In accordance with the intimation contained in the last clause of the report of the "Joint Special Committee on Celebration of the successful Laying of the Atlantic Telegraph Cable," the following extended report of the proceedings had in that matter, together with the various remarks and interesting details connected therewith, is respectfully submitted to the Honorable the Common Council of the city of New York, by

CHARLES T. McCLENACHAN,

*Secretary to the Joint Special Committee on Celebration, &c.*

New York, December 25, 1858.

# REPORT.

About noon on the 5th of August, 1858, intelligence was received in the city of New York of the successful accomplishment of an enterprise in which the deepest interest had been manifested by all classes of the community—namely, the submerging of the Electric Cable for telegraphic purposes between the continents of Europe and America.

The first announcement was made by Mr. Cyrus W. Field to the Mayor of New York City in the following words:

"Trinity Bay, August 5, 1858.

"*Mayor of New York:*

"Sir—The Atlantic Telegraph Cable has been successfully laid.

"C. W. FIELD."

To which his Honor replied:

"Mayor's Office,
"New York, August 6, 1858.

"*To Cyrus W. Field, Esq., Trinity Bay:*

"Sir—Your dispatch has been received. I congratulate you for myself and for the people of this city on the success of the great work of uniting together the Old and the New World by the electric telegraph. Science, skill and perseverance have finally triumphed.

"DANIEL F. TIEMANN,
"*Mayor of New York.*"

The following telegraphic dispatch was received on the same day, and given to the public through the medium of the various journals:

DISPATCH FROM MR. CYRUS W. FIELD.

"TRINITY BAY, August 5, 1858.

"*To the Associated Press, New York:*

"The Atlantic Telegraph fleet sailed from Queenstown on Saturday, July 17; met at mid-ocean on Wednesday, the 28th, and made the splice at 1, P. M., on Thursday, the 29th, and then separated—the Agamemnon and Valorous bound to Valentia, Ireland, and the Niagara and Gorgon for this place, where they arrived yesterday, and this morning the end of the Cable will be landed.

"It is sixteen hundred and ninety-eight nautical, or nineteen hundred and fifty statute, miles from the Telegraph house, at the head of Valentia Harbor, to the Telegraph house, Bay of Bull's Arm, Trinity Bay; and for more than two-thirds of this distance the water is over two miles in depth.

"The Cable has been paid out from the Agamemnon at about the same speed as from the Niagara.

"The electrical signals sent and received through the whole Cable are perfect.

"The machinery for paying out the Cable worked in the most satisfactory manner, and was not stopped for a single moment from the time the splice was made until we arrived here.

"Captain Hudson, Messrs. Everett and Woodhouse, the engineers, the electricians and officers of the ships, and, in fact, every man on board the Telegraph fleet, has exerted himself to the utmost to make the expedition successful;

and, by the blessing of Divine Providence, it has succeeded.

"After the end of the Cable is landed and connected with the land line of telegraph, and the Niagara has discharged some cargo belonging to the Telegraph Company, she will go to St. John's for coal, and then proceed at once to New York.

"CYRUS W. FIELD."

A supplementary dispatch, announcing that the Agamemnon had been as fortunate as the Niagara, was also received the same day by the New York Herald, from its special correspondent on board the latter ship. It ran as follows:

"UNITED STATES FRIGATE NIAGARA,
"TRINITY BAY, N. F.,
"August 5, 1858.

"*James Gordon Bennett, Esq.:*

"The Atlantic Telegraph Cable is laid! The United States frigate Niagara has reached Trinity Bay, Newfoundland, and landed her end of the line. The Agamemnon reached Valentia, Ireland, on the same day, with her end of the Cable.

"The electrical communication is perfect. Signals are passing between the two stations with the greatest accuracy.

"The Niagara will leave for St. John's, N. F., on the 5th instant, and will start for New York after taking on board sufficient coal for the passage.

"The laying of the Cable was accomplished by both ships in a little over six days."

The latter dispatch was so confirmatory of the complete

success of the enterprise, that there was no longer any room for doubt; and, consequently, the manifestations of rejoicing over the event were witnessed in every quarter, not alone in the city of New York, but throughout all portions of the United States to which the intelligence had penetrated.

The President of the United States conveyed to Mr. Field his sense of the importance of the event in the following dispatch:

"BEDFORD, Penn., August 5, 1858.

"*To Cyrus W. Field, Trinity Bay:*

"MY DEAR SIR—I congratulate you with all my heart on the success of the great enterprise with which your name is so honorably connected. Under the blessings of Divine Providence I trust it may prove instrumental in promoting perpetual peace and friendship between the kindred nations.

"I have not yet received the Queen's dispatch.

"Yours, very respectfully,

"JAMES BUCHANAN,"

The communication from Captain Hudson, of the Niagara, to his family, was in these words:

"TRINITY BAY, August 5, 1858.

"God has been with us. The Telegraph Cable is laid without accident, and to Him be all the glory. We are all well.

"Yours, affectionately,

"WM. L. HUDSON."

On the 11th of August, the Joint Special Committee appointed by the Common Council to do honor to the occasion, entered upon the arduous, but pleasant duty of making such arrangements or celebrating the great event thus announced, in such a manner as should testify the high appreciation in which it was held by this community.

Alderman McSpedon, was unanimously chosen Chairman, and Charles T. McClenachan and Francis J. Twomey were appointed to act as Secretaries.

The Committee having, after deliberation, concluded that it would be advisable to divide the celebration into two parts, as, owing to the impossibility of the principal actors in the great event to be in New York by the time that the messages between the Queen of Great Britain and the President of the United States should pass over the wire, the honors intended could not be done to them at the same time that that event should be celebrated, at once made arrangements for a grand display of fireworks at the City Hall, and for an illumination (in which latter all who could were invited to join), whenever it should be announced that the friendly greetings had passed between the heads of the two great nations.

The Message of the Queen of England was received on the 16th of August, and was as follows:

" *To the President of the United States, Washington:*

" The Queen desires to congratulate the President upon the successful completion of this great international work, in which the Queen has taken the deepest interest.

" The Queen is convinced that the President will join with her in fervently hoping that the Electric Cable which now connects Great Britain with the United States will prove an additional link between the nations, whose friendship is founded upon their common interest and reciprocal esteem.

"The Queen has much pleasure in thus communicating with the President, and renewing to him her wishes for the prosperity of the United States."

To this greeting, the President of the United States sent the following reply:

"WASHINGTON CITY, August 16, 1858.

"*To Her Majesty Victoria, Queen of Great Britain:*

"The President cordially reciprocates the congratulations of her Majesty the Queen, on the success of the great international enterprise accomplished by the science, skill and indomitable energy of the two countries.

"It is a triumph more glorious, because far more useful to mankind, than was ever won by conqueror on the field of battle.

"May the Atlantic Telegraph, under the blessing of Heaven, prove to be a bond of perpetual peace and friendship between the kindred nations, and an instrument destined by Divine Providence to diffuse religion, civilization, liberty and law throughout the world.

"In this view, will not all nations of Christendom spontaneously unite in the declaration that it shall be forever neutral, and that its communications shall be held sacred in passing to their places of destination, even in the midst of hostilities.

"JAMES BUCHANAN."

In commemoration of the passage of these communications through the Atlantic Telegraph Cable, the City Hall and the public buildings of New York were brilliantly illuminated, and a magnificent display of fireworks, prepared by Mr. J. Edge, Jr., was exhibited in front of the City Hall on the evening of the 17th August.

After the exhibition of fireworks was over, the Committee proceeded, with a band, to the residence of C. W. Field, Esq., and serenaded Mrs. Field, and subsequently serenaded Mr. Peter Cooper, President of the New York, Newfoundland and London Telegraph Company, his Honor Mayor Tiemann, and the British Vice-Consul Mr. Archibald.

It was the subject of much regret that, early on the ensuing morning, the dome of the City Hall was consumed by fire, caused, doubtless, by sparks from the fireworks having entered under some defect in the metal covering of the roof.

A Committee, appointed to investigate into the cause of the disaster, made the following report:

Your Committee, having inquired fully into the action of the Sub-Committee on Illumination, as directed by your resolution, respectfully report:

That the Sub-Committee, finding it impossible to obtain a supply of water through hose from the Croton pipes, there not being sufficient head to carry the same upon the roof of the Hall, caused to be procured a large number of buckets, which they had filled with water, carried upon the roof and used to wet the same and the wood-work about the cupola between each successive display of fire-works; they also requested that a fire-engine be stationed in the vicinity of the City Hall, which was done, and the engine remained where stationed until some time after the conclusion of the display; and that directly after the conclusion of the exhibition they caused the roof and wood-work to be examined and wet, and all the fragments found on or about the roof gathered together and extinguished that at 9½ o'clock the Chairman of the Joint Committee, with the Sub-Committee made a minute and particular examinaton of the Hall, and were informed by Mr. Edge that he had used all the precautions they had directed, and others of his own suggestion; that at the request of the Committee, Mr. Edge again made an examination of the roof and cupola of the building, and reported that, so far as he could discover, everything was right. The Sub-Committee, with many members of your Committee, remained at the Hall until after 10 P. M., and did not leave until the Superintendent of Repairs and Supplies and his Deputy had left, believing that all was safe.

The Sub-Committee caused to be left upon the roof a number of buckets of water to extinguish any undiscovered remains of the fireworks.

Your Committee are therefore of opinion that every possible precaution was used by the Sub-Committee, and that more than usual care was taken by them and Mr. Edge, and they are of the opinion that had the same care and precaution been exercised by those in whose charge the Hall was left by them, the unfortunate accident would not have occurred.

Your Committee suggest that a thorough examination be had into the origin of the fire, and submit the following resolution:

*Resolved,* That the Mayor and Fire Marshal be requested to thoroughly investigate and report to the Common Council the origin of the fire, and to the negligence of what person or persons the same is to be attributed.

J. LYNES, } *Sub-Com. of In-*
S. A. BUNCE, } *vestigation, &c.*

The Niagara arrived off the Battery at 4 o'clock of the afternoon of Wednesday, the 18th, and was saluted by the guns of all the forts and the shipping, as well as by sections—a battery served by detachments of the Scott Life Guard, who had been detailed for that duty by Commissary-General J. H. Hobart Ward, in accordance with the following letter and order:

HEADQUARTERS SCOTT LIFE GUARD,
NEW YORK, Aug. 13, 1858.

*C. T. McClenachan, Esq.*,

DEAR SIR—Inclosed I have the honor to transmit a copy of an order of this Corps, directing the firing of the salutes by detachments from the Scott Life Guard, pursuant to the arrangements determined upon by the Joint Committee of the Common Council on Celebration.

Permit me to add that this Corps, participating in the spirit which animates the Common Council as well as the country at large, have offered to perform this service as expressive of their gratification at the successful issue of the great undertaking so happily concluded.

Most respectfully,
J. H. HOBART WARD,
*Captain Commanding.*

HEADQUARTERS SCOTT LIFE GUARD,
NEW YORK, Aug. 13, 1858.

ORDER NO. —

The municipal authorities of this city having determined to salute the United States steamship Niagara and her gallant Captain, Hudson, on their arrival at this port, after the successful laying of the Atlantic Cable, by firing a salute of 100 guns at the Park, and the same number at the Battery, it is thereupon ordered, at the request of the

Mayor and Joint Committee of the Common Council on Celebration, that this Corps furnish the necessary detail to fire such salutes; and the following detail is hereby made: *Detail for Park*—Lieut. Wm. H. Browne, and eight men. *Detail for Battery*—Lieut. J. D. MacGregor and eight men.

The First Sergeant will make the above detail of men with all possible dispatch, designating the men for each command, with directions to report to their respective commanders, at the Armory of the Corps, at 11, A. M., on Tuesday 17th inst., in full uniform except cartridge-box.

By order of J. H. HOBART WARD,

*Captain Commanding.*

W. B. Parisen, *First Sergeant.*

The following were the principal sub-committees named to make the necessary preparations and arrangements for the grand demonstration which it was decided should be made on the 1st September, and for the municipal dinner, that should be given in honor of the participants in the accomplishment of the great enterprise just achieved, on the ensuing day:

*On Fireworks*—Alderman Boole and Councilmen Dunn and Bunce.

*On Programme*—the Chairman, and Mr. R. W. Lowber.

*On Procession*—Aldermen Lynes and Owens.

*On Decorative Arches over Gates to the Park*—The Deputy Superintendent of Repairs and Supplies and Councilman Bunce.

*On Carriages*—Aldermen Lynes and Boole, and Councilman Mulligan.

*On the Grand Public Dinner and Decorations at the Crystal Palace*—Councilmen Bunce, Ross and Bickford, and Aldermen Lynes and Hoffmire.

Upon request, the following named gentlemen kindly consented to lend their valuable services in assisting to make the celebration all that could be desired:

The Hon. Edward Everett, in preparing the address to Captain Preedy and the officers of H. B. M. steamers Gorgon and Valorous, upon the presentation of testimonials at the Crystal Palace;

The Right Reverend Archbishop Hughes, in preparing the address to be delivered to the representatives of the "New York, Newfoundland and London," and "Atlantic" Telegraph Companies;

Prof. Joel T. Headley, in preparing that to Captain W. L. Hudson and officers and men of the Niagara;

Prof. Isaac Ferris, in preparing that complimentary to Engineers Everet and Woodhouse, and the electricians of the expedition;

Hon. George Bancroft, in preparing that to be presented to Captain Dayman of H. B. M. Steamer Gorgon, which had served as tender and guide to the Niagara.

D. D. Field, Esq., aquiesced in the desire that he would assume the important position of "orator of the day," and the Rev. Drs. Mott and Field accepted the invitation to open and close the ceremonies at the Crystal Palace with prayer and a benediction.

Rev. Dr. Ogilvie also signified his readiness to comply with the request that he would pronounce "Grace" upon the occasion of the contemplated grand municipal dinner to be given at the Metropolitan Hotel.

A communication from Commodore Kearney stated that Commander Ward had kindly placed the splendid band of the Ship-of-the-line North Carolina at the service of the Committee.

From many others, proffers of services were received and accepted, of which due mention is made in the following pages.

Brigadier-General Charles Yates, having been nominated Grand Marshal of the day, for the 1st of September, accepted that arduous post, and to his excellent management was, in a great measure, owing the very happy manner in which the affair passed off.

A special meeting of the engineers and foremen of the Fire Department had been held on the 13th of August, to make arrangements for the celebration of the transmission of the expected message from the Queen of England to the President of the United States, and the President's reply; when the following resolutions were adopted, on motion of Mr. J. H. Carolan, of Hook and Ladder Company No. 12:

*Whereas*, extensive arrangements are being made, both in this city and throughout the entire country, to celebrate, in a becoming manner, the success of the "great event of the age," the joining of the Old and New Worlds by means of the Atlantic Cable; and whereas, this wonderful triumph of man's genius, industry and energy over all obstacles of time, distance or the vast deep, which has sent a thrill of joy throughout the entire length and breadth of the land, calls for some public expression from our Department; therefore, be it

*Resolved,* That, the Fire Department of the city of New

York unite with the municipal authorities and our fellow-citizens in their proposed demonstration in commemoration of this wonderful event by having a grand torch-light parade on the evening of the —— inst.

*Resolved*, That no Companies be allowed to parade more men than they are allowed by law, as follows:—First class engines, seventy men; second class, sixty men; third class, fifty men; hose companies, thirty men; hook and ladder companies, fifty men.

*Resolved*, That the Chief Engineer act as Grand Marshal on the evening of the parade, and that the Board of Engineers be empowered to make necessary arrangements for the same.

*Resolved*, That the Board of Fire Commissioners, the Board of Trustees, the Exempt Firemen and Assistant Engineers be, and are hereby invited to participate with us in the parade.

It was subsequently concluded, however, that the grand torch-light procession should take place on the day of the general celebration.

The New York Chamber of Commerce also held meetings, to take measures for extending some proper mark of respect to Captain Hudson and the officers of the United States Steamship Niagara, and appointed the following Committee to consider and recommend the best mode of acknowledgment:

| | |
|---|---|
| William B. Astor, | George W. Lane, |
| Lloyd Aspinwall, | Richard Lathers, |
| Wm. H. Aspinwall, | Marshall Lefferts, |
| Samuel D. Babcock, | A. A. Low, |

Caleb Barstow,
A. Belmont,
James Brown,
George W. Blunt,
Edwin J. Brown,
James H. Brower,
James H. Brundage, Jr.,
Leopold Bierwith,
Hugh W. Camp,
Horace B. Clafflin,
Frederick A. Conklin,
Simeon B. Chittenden,
James B. Colgate,
Denning Duer,
W. Butler Duncan,
W. W. De Forrest,
George Folsom,
Benjamin H. Field,
Robert C. Goodhue,
George Griswold, Jr.,
Walter S. Griffiths,
Wilson G. Hunt,
William F. Havemeyer,
Charles A. Heckscher,
George Hall,
Edwin Hoyt,
Abraham S. Hewitt,
Edward Hincken,
John D. Jones,
James Lee,
F. S. Lathrop,
Robert B. Minturn,
James M. Morrison,
E. D. Morgan,
Charles H. Marshall,
Anthony B. Neilson,
J. De Peyster Ogden,
Pelatiah Perit,
John K. Myers,
George T. Olyphant,
Henry E. Pierrepont,
Augustus C. Richards,
George S. Robbins,
John A. Stevens,
James E. Southworth,
Oliver Slate,
Samuel B. Shepherd,
William K. Strong,
Augustus E. Silliman,
T. B. Satterthwait,
Henry A. Smythe,
Daniel D. Smith,
Henry M. Schiefferlin,
James S. T. Stranahan,
Moses Taylor,
Thomas Tileston,
Ellwood Walter,
William Watson,
John D. Wolfe,
Luther B. Wyman.

All the military, civil and other societies also held meetings for the purpose of assisting in the celebration, and were largely represented in the procession on the 1st of September.

Subsequently to the transmission of the messages between the Queen of Great Britain and the President of the United States, the following dispatches of congratulation passed between the Mayor of New York and the Lord Mayor of London:

MAYOR'S OFFICE, NEW YORK,
August 10, 1858.

*To the Right Honorable Sir Robert Walter Carden, M.P., Lord Mayor of London:*

I congratulate your Lordship upon the successful laying of the Atlantic Cable, uniting the continents of Europe and America and the cities of New York and London—the work of Great Britain and the United States—the triumph of science and energy over time and space, thus uniting more closely the bonds of peace and commercial prosperity, and introducing an era into the world's history pregnant with results beyond the conception of a finite mind. To God be all the praise.

D. F. TIEMANN,
*Mayor of New York City.*

LONDON, August 23, 1858.

*From the Lord Mayor of London to the Hon. D. F. Tiemann, Mayor of New York:*

The Lord Mayor of London most cordially reciprocates the congratulations of the Mayor of New York upon the success of so important an undertaking as the completion of the Atlantic Telegraph Cable.

It is, indeed, one of the most glorious triumphs of the age, and reflects the highest credit upon the energy, skill and perseverance of all parties intrusted with so difficult a duty, and the Lord Mayor sincerely hopes that, through the blessing of Almighty God, it may be the means of cementing the kind feelings which now exist between the two countries.

R. W. CARDEN,

*Lord Mayor of London.*

That the rejoicings, which there was no doubt would be celebrated throughout the United States and British provinces, should be more appropriate, the committee instructed the secretary to communicate to the different cities, &c., their request that they should be so timed as to be simultaneous with those to take place in New York; and accordingly the following invitation was published, and the substance thereof transmitted by telegraph to the more distant cities and towns:

"The Joint Committee of the Common Council of the city of New York on celebrating the laying of the Atlantic Telegraph Cable, having determined to celebrate that event by a salvo of one hundred guns and national salutes, with ringing of all the church bells, immediately upon the transmission of the reply of the President of the United States to the message of her Majesty Queen Victoria, and with bonfires and illuminations on that day, and also by a military and civic procession some days thereafter, to be followed by a municipal dinner, which festivities are understood and expected to take place simultaneously with those in the city of London and throughout Great Britain and the Continent, would request the citizens throughout the United States and British provinces to unite with them and order their festivities to take place at the same time and in the same order, of which due notice will be given.

"By order cf the Committee,

"C. T. McCLENACHAN, *Secretary.*"

So far as heard from, this request was complied with, so that the celebration of this glorious consummation has been more general and unanimous than any ever before known in the civilized world.

I would now beg leave to submit a description of the proceedings and festivities of the first and second of September, which are somewhat minutely given, as they will form, hereafter, a portion of a period in history more memorable than can be presented by any preceding age—for, if no lasting benefit may be derived from the Cable just laid, it has been clearly demonstrated that the enterprise was feasible, and some of the most vexed questions in relation to electro-telegraphy have been solved, while there is no doubt that this achievement will give an impetus and confidence to energy and enterprise that will ultimately result in what we so fondly hoped from the present attempt.

## THE CELEBRATION OF THE FIRST OF SEPTEMBER—THE PUBLIC DECORATIONS.

The morning of the 1st of September, the day set apart for the *fête*, broke fair and bright. The heavens were smiling and propitious. An early September sun, shorn of the fierceness of his beams by the bracing gale that wafted over the city, cast his genial glow on all around and underneath. The gray dawn was ushered in with the echoing sounds of the hammer and the saw and all the busy din of preparation. As the morning advanced, the streets assumed a holiday look, banners, flags, mottoes, meeting the eye at every step, and eliciting the admiration of the restless hosts which crowded the sidewalks, thronged the corners, and rushed into Broadway, the grand centre of attraction, from every street and avenue. All the approaches to the metropolis poured in constant streams of human beings. The ferry-boats, though running with all possible frequency, failed utterly to supply the demands made upon them by residents and visitors in the suburbs. The early, and, indeed, the late trains over the New Jersey, Long Island and northern roads were crammed past all precedent. The carriage-ways were one long cloud of dust from the ferry-houses miles out into the country in all directions.

The shipping in the harbor, lavishly bedecked with bunting, suggested the idea of a forest of flowers. The East River and the North, for miles and miles, looked like a crushed rainbow—an *entourage* of confused iridescences, inclosing a panorama of gorgeous thoroughfares and rejoicing people.

The suspension of business was almost universal. The principal stores closed at an early hour in the forenoon. Nearly all were closed at mid-day. The banks if they opened at all remained so for but an hour or two. The Government offices were shut altogether, and the Courts were completely deserted. The decorations of the City Hall, and the inscriptions displayed on Broadway, are deserving of notice.

From all the flagstaffs on the Hall gay banners and streamers were flying. Surmounting the centre of the roof was a painting representing the national bird with wings outspread, standing on a starred and striped shield, as if just lighted there, with the lion and the American standard under his right pinion, and the English standard and unicorn under his left. Under this was the name of FRANKLIN in large letters, on one side, the coat of arms of the State of New York, and on the other, the coat of arms of the city of London.

On the west side were banners with the following:

"FIELD, WOODHOUSE, DAYMAN, COOPER, HUDSON."

And on the east.

"PREEDY, GURNEY, ALDHAM, EVERETT, MORSE."

Facing Broadway were the following:

"MAURY, WHITEHOUSE."

And facing Centre street:

"BRIGHT, BERRYMAN."

The east gate of the Park, was surmounted by an immense banner, formed like an arch, bearing on one side the words:

"THE PATH OF FRANKLIN LED TO A FIELD OF ENTERPRISE."

At one end of the inscription appeared an old-fashioned printing press, at the other, a telegraphic apparatus.

On the other side were displayed the words:

"PEACE HATH HER VICTORIES
NOT LESS RENOWNED THAN WAR."

Similar arches were erected at the western and southern gates. That at the western gate bore on one side the inscription:

"THE NIAGARA AND AGAMEMNON, WITH THEIR ESCORTS GORGON AND VALOROUS—THE FOUR VICTORS IN THE GREAT BATTLE BETWEEN SCIENCE AND NEPTUNE."

At one end was the figure of an anchor, wreathed with flowers, and above it, the British crown. The other side exhibited the words and motto :

"FIELD AND EVERETT, HUDSON AND PREEDY, DAYMAN AND ALDHAM. 'NAMES GREAT IN MOUTHS OF WISEST CENSURE.' "

At the upper end of this was represented another anchor, surmounted by the American eagle.

The arch at the southern gate had on one side the following :

"THE TRINITY MARRIAGE—UNITED IN THE BONDS OF PEACE AND AMITY, AUGUST 5, 1858, ALBION TO COLUMBIA, HUDSON TO HISTORY, FIELD TO FAME."

Underneath this was the inscription :

"WHOM GOD HATH JOINED TOGETHER
LET NO MAN PUT ASUNDER."

At one end was the shield of Great Britain and the word "Victoria," and at the other, the shield of the United States and the word "Buchanan." On the other side was the following :

"THERE IS A PATH WHICH NO FOWL KNOWETH, AND WHICH THE VULTURE'S EYE HATH NOT SEEN ; THE LION'S WHELPS HAVE NOT TRODDEN IT, NOR THE FIERCE LION PASSED BY IT."—*Job*, xxviii., 7, 8.

On Broadway, commencing at the Battery and proceeding upwards, the following inscriptions and designs were observable on either hand.

A large transparency in front of No. 118, bore the following:

"To lay the Ocean Cable
The Yankees proved quite able,
John Bull was not behind
Valentia Bay to find,
And thus the thing was done
By John and Jonathan."

Next door (120) exhibited as an emblem, an

All-seeing Eye,
Surrounded by a glory, and looking down upon
two hands holding a laurel wreath over
The Niagara, bearing the English flag. [A GLOBE] The Agamemnon, bearing the American flag.

FRANKLIN. American Continent. European Continent. MORSE.

At the store corner of Fulton street, and east side of Broadway:

"AMERICAN ENTERPRISE.

There were Kings before Agamemnon, but the Agamemnon is King of them all.

—

It took two American rivers to conquer the Atlantic:

THE NIAGARA AND HUDSON.

Morse, the inventor, supplied the germ;
Steers, the mechanic, supplied the Niagara;
Field, the business man, completed the glorious work.
Americans exult!
American invention, mechanical genius, and business energy and perseverance—furnished by Morse, Steers, and Field—do honor to America.

THE UNION OF
ENGLAND AND AMERICA.

In place of the wedding
Ring, the Atlantic Cable."

On a transparency in Fulton street, appeared the following squib:

"SPECIAL DISPATCH BY THE ATLANTIC CABLE.

Ballyragget, Ireland.
September 1, 1858—3½ o'clock, P. M.

To WM. EGLINGTON, 140 Fulton street, New York.

Faith, Bill, I'm very glad to hear that you are well, and able
To send a graphic sketch to say so through the cable.
No cows here have got the "murrain," no sheep have got the rot;
Potatoes they do flourish now, throughout the bright green spot.
I like to hear from all of ye, but I'd like to see you better;
Send back a toast by "cablegraph," and the liquor in a letter.
But wont't it be a glorious thing, when science will be able,
To let us hear your own sweet "brogue" by electrographic cable.

*Dum vivimus bibamus*—(While we live let us drink.)

CHAS. COLTER.

ANSWER.

For your telegraph dispatch I paid cheerfully $20.

W. E."

Barnum's Museum, the Express offices, and all the hotels between the Battery and Fulton street, and many of the stores and restaurants, were festooned with the flags of the two nations.

From the top of the Astor House the English and American flags were flying, and in front, over the entrance, were placards bearing the following inscriptions:

"The Age of Progress, 1858."
"Lightning under the Ocean."
"Steam on the Erie Canal."

"The Atlantic Telegraph transmits the lightning of Heaven, and binds together 60,000,000 human beings speaking the same language, and worshiping the same God."

The establishment No. 247 Broadway was brilliantly adorned.

From the beak of the large golden eagle which stretched over the top of the building hung a portion of the Telegraphic Cable, one side of which was looped to the American, and the other to the English flag. At one end of the Cable was a transparency representing the U. S. steam frigate Niagara, and at the other the British steam frigate Agamemnon. Above the eagle, and covering him, was an arched transparency, which bore the legend

"THE ATLANTIC TELEGRAPH CABLE."

and below this was another transparency, inscribed with the following:

"PROJECTED IN 1854.<br>"SUCCESSFULLY LANDED AUGUST, 1858."

Another transparency was displayed below this, and between the windows, on which was inscribed the name of

CYRUS W. FIELD,

and over the front door, extending along the whole front of this magnificent edifice, was a transparency as follows:

MALE AND FEMALE.

| | | |
|---|---|---|
| ENGLAND. | Hands in a cordial grasp. | AMERICA. |

In front of the large concern Nos. 258, 259 and 260, Broadway, and extending the whole length of the building, was a transparency after this fashion:

| American Flag. | A streak of lighting connecting both flags,<br>"Thunder and Lightning."<br>The Telegraph's Lightning<br>is<br>"Our Thunder." | English Flag. |
|---|---|---|

The store on the north-west corner of Broadway and

Warren street was decorated with flags of England, Scotland, Ireland and America, and was brilliantly illuminated at night. The only transparency was one bearing the rose, thistle and shamrock, with the word "Union" between them.

The establishment on the north-east corner of Broadway and Pearl street was brilliant with small flags of all nations. In the windows were the following inscriptions: "America to all nations. E Pluribus Unum." "Europe and America. Blest be the tie that binds." "New York and London one city." "Niagara and Agamemnon. Hudson and Preedy." "Directors of the Atlantic Telegraph Company. Their monument lies low." "The Atlantic Cable, the Necklace of the World."

Over the clock store, No. 338 Broadway, "Time works Wonders" was placarded.

At No. 343 Broadway was the following:

"Honor to those whose genius led
The lightning's track through ocean's bed,
A path for thought;
And kindred honors may they share
Who sweet relief from toilsome care
For Woman wrought."

This building was gay with flags of various nations. On the right of the transparency was written the word "Past,' and on either side of it were representations of a courier on horseback, a mower mowing, a woman toiling at her needle, a canoe propelled by hand, and a hand and pen writing; on the left side was the word "Present," with a mowing and reaping machine, Telegraph wires, a woman working with a sewing machine, a printing press and a steamboat.

At No. 345 Broadway was the following inscription :

Through ocean's desolation deep
Thoughts in "electric fire" sweep—
The "Friendly Nations" converse hold
Through its caverns dark and old.

"Noble theme," in waves of song
Roll triumphant notes along—
"Fairy's laurel" decks a *Saxon* brow—
Our Country's FIELD is victor now.

A large painting covered two stories of the front of No. 349 Broadway. It represented an eagle on one side and a lion on the other, with the Cable stretching through the sea between them. In the centre, on a rock, was a female form with outspread wings dominant over commerce and the arts. Above, on either side, was written, "VICTORIA," "BUCHANAN." Below was this inscription: "CYRUS W. FIELD, DAGUERRE and MORSE. The one harnessed the light, the other the lightning. May the 'continuity' of the Cable be as perpetual as the rays of the sun." To the left of this was, "Captain PREEDY, Agamemnon," and to the right, "Captain HUDSON, Niagara."

From the gilded eagle that ornaments the fourth story of the establishment No. 356 Broadway, streamers hung. From the first floor a temporary balcony projected with a front of canvass, rows of upright muskets with candles stuck in the barrels, and the couplet,

"The Cable with its peaceful tricks
Makes of muskets candlesticks."

The whole front of No. 359 Broadway, the lower part of which is used as a dining saloon and the upper stories as a photographic gallery, was decorated with a splendid transparency fifty by twenty-five feet. It bore on the top the words "Science, Labor and Art—Union Cable."

On the right was a portrait of FIELD, in the centre one

of Franklin, and on the left one of Morse. Beneath were two female figures, representing America and England joining hands; while to the right of these, and below, was a steamship, with the emblems of mechanism and science; and to the left, was a sailor from the Niagara. Under this was a bridge joining two countries, and beneath this again was a figure of Neptune chained, implying that his power was at an end. To the right was the letter B, and to the left the letter V, both surrounded by electric stars. The whole design was well conceived and beautifully executed. It attracted a great deal of attention.

Over the front of Nos. 365 and 367 was this inscription:

VICTORIA.

AGAMEMNON.

All honor to the inventive genius and indefatigable enterprise of

John and Jonathan,

that has succeeded in consummating the mightiest work of the age. May the cord that binds them in the bonds of

International

Friendship never be severed, and the Field of its usefulness extend to every part of the earth.

—o—

Let nations' shouts, 'midst cannons' roar,
Proclaim the event from shore to shore.

NIAGARA.

BUCHANAN.

No. 369 bore a huge painting, representing the laying of the Cable, with the letters "M." "F." "H." at the top, and beneath, "Europe and America united by telegraph," "Glory to God in the highest, on earth peace, good-will toward men," "Buchanan," "Victoria."

Florence Hotel, corner of Walker Street and Broadway, was decorated with colored lanterns, and flags, and small transparencies, bearing the names of the distinguished men who were connected with the submarine telegraph enterprise.

The front of Nos. 374 and 376 Broadway was tastefully arranged with the flags and emblems of England and America during the day. There were large transparencies.

On the central one, headed

"THE ANGLO-SAXON TWINS,"

were two figures representing England and America, in the persons of John Bull and Brother Jonathan, the Cable uniting their two hearts. Underneath was the following verse:

"The brothers need no bulwarks,
No towers along the steep;
Our march is on the mountain wave,
Our lightning through the deep."

On the left was another large transparency, representing two first-class ships of war engaged in deadly strife. Under the sea of blood are represented, not the glories, but the realities of war—hundreds of human beings were seen sinking as a chain of death. The background represented the Demon of Destruction. This picture was simply described as the "Past."

To the right was a companion picture called the "Present." This was represented by the Niagara and the Agamemnon engaged in the peaceful laying of the Cable, emanating from which sparkles, in electric fire, the words—"Peace on earth, to men good-will." The background represented the hand of God assisting in this wonderful work—with the words of Captain Hudson: "God has been with us."

No. 424 Broadway had a fine transparency. The letters "B," "V," were at each of the upper corners; at the bottom corners were the English and American shields. In the middle was the following:

"Morse, Field and Hudson—three noble mates,
Have made all nations the United States."

Below this were the Niagara and Agamemnon laying the Cable:

> "Europe from these United States
> The sea no longer separates."

No. 445 Broadway displayed on the top of a transparency "XIX Century;" on one side, the names of C. W. FIELD, HUDSON, DAYMAN and EVERETT; on the other side, the names of WATT, FULTON, STEPHENSON and MORSE. Below, the Cable, resting on New York and London, with the word "United" over it, with the Niagara and Agamemnon in the distance.

The City Assembly Rooms, which were hired and decorated by the Young Men's Democratic Union Club for the occasion, presented an imposing appearance. They were profusely decorated with the flags of all nations. Outside were three pedestals, that on the right bearing a statue of Morse, that on the centre one of Franklin, and that on the left one of Field. In the centre, extending along the whole front, was a piece of canvas, on which was written: "The Atlantic Telegraph: An artery through which the living blood of Liberty shall be infused through all nations over the sea."

On the new building corner of Grand and Broadway, was a transparency with a female figure on each end. Britannia with the English flag on the right, and an Indian woman with the American flag on the left. Above was the following motto:

> "The Great Event of our Day—Commenced in Faith, pursued in Hope; may it be a bond of Charity and Good-Will between Nations."

Under the above inscription were the Niagara and Agamemnon laying the Cable, and between them a shield bearing the arms of England and America, surmounted by a "Spread Eagle."

The exterior of the establishment on the north-east corner of Broadway and Broome streets, was gorgeously decorated.

That of 497 Broadway had this inscription:

"Married August, 1858, Europe to America. They were married by lightning, and "by thunder" they shall never be divorced."

The St. Nicholas Hotel decorations were among the finest on Broadway. The roof was ornamented with three large flags—two American and one British. Suspended across the street were the American, French and Spanish ensigns, and immediately under them floated a banner bearing the following inscription:

"A portion of the first submarine Cable laid by Col. Sam. Colt, in 1843, from New York to Coney Island, via Hell Gate. Man proposes—God disposes."

Three fine transparencies ornamented the marble front of this building. The centre piece represented Buchanan and Victoria shaking hands, the Niagara and Agamemnon being seen in the distance. On each side of these figures hung the flags of the United States and Great Britain, and underneath the entire was the single word:

"RECIPROCITY."

The second transparency represented John Bull and Brother Jonathan united by means of the Atlantic Cable. Pillars on each side of the figures, on which were wreathed the names of Morse, Hudson, Maury and Everett on one side, and Franklin, Steers, Field and Cooper, on the other. Beneath was a figure of Morse, holding an umbrella in his hands and treading upon lightning, surmounted with the following inscription:

"MORSE, THE LIGHTNING TAMER."

The third transparency embraced a number of figures,

representing Puck putting a girdle round the earth, Morse killing time with lightning speed, and the Queen and Buchanan performing a new feat on the slack rope. Clustered around these various figures were the following inscriptions:

"TELEGRAPH LINES ARE THE NERVES OF NATIONS.

UNITED FOREVER.

PEACE BLESSES THE UNION."

At 555 Broadway, the following was seen:

"REJOICE ALL NATIONS.

MARRIED,

On Thursday, August 5, 1858,
In the Church of Progress,
At the altar of Commerce,
The Old to the New World.
May they never be divorced!

BORN,

On Monday, August 16, 1858,
In the bed of the ocean,
Of science and enterprise,
The child Atlantic Telegraph.
May it live to honor its parents!

DIED,

Monday, August 16, 1858,
From an electric shock,
OLD FOGYISM.
May he rest in peace!
ejoice at the marriage, the birth, and the death."
[The American and British flags intertwined.]

The St. John's House displayed a motto: "Genius and Perseverance have caused both ends to meet."

The Metropolitan Hotel, over the front of the balcony of the portico had hung in folds the flags of several European powers. The Union Jack was placed on the centre of the roof, and at each end the American ensign. The whole front was decorated with a magnificent display of

Chinese lanterns. The building is five stories, with thirty windows on each floor. Sixty cylindrical lanterns were run on immediately below the edge of the roof, thirty spherical lanterns in the face of the front, and a corresponding row of cylindrical lanterns over the plinth of the second floor windows, the general effect, even by daylight, being very picturesque.

Niblo's Garden had over the portico the Union Jack and American flag with the lyre of Terpsichore in dead white ; transparency inscribed, "When the multitude saw it, they marveled and glorified God, that such power was given unto men."

Laura Keene's Theatre, was tastefully decorated with the flags of all nations. In front were models of the Niagara and Gorgon, working by machinery and firing salutes at appropriate intervals in the passage of the procession. Outside, was a large transparency, representing a stage with footlights, scenery, &c., &c., bearing the motto :

"There's no such thing as fail for Saxon blood,
Wherever pours the stream, by FIELD or flood."

From the City Guard Room to the Egyptian Museum a banner suspended across the street on which were the United States and British flags with the following :

"NEW YORK CITY GUARDS.
*Ratione aut vi.*
These are the banners from whose folds unfurl'd
Fair freedom flings her blessings o'er the world."

Burton's Theatre was decorated with flags, under which was a transparency, showing the Cable line from "America" to "England," surmounted by the American eagle bearing the olive branch of peace in one talon and the forked lightning in the other. Under which was the following:

"Smile, Heaven, upon thisfair conjunction! Enrich the time to come with smooth-faced peace, with joyous plenty, and fair prosperous days."

At Lafarge House:

"Science controls the elements; it binds together with an iron cord the ends of the earth. Guided by wisdom, man unites his thoughts with the lightning's point."

Underneath, a transparency of ships at war, and a vessel at sea in a state of explosion, with the words:

"SEVERED JULY 4, 1776.
UNITED AUGUST 12, 1858."

The Hone House had a banner with the following:

| "SCIENCE, MORSE. | MECHANISM, EVERETT. | PERSEVERANCE, FIELD." |
|---|---|---|

At 683 Broadway were the following:

"THE CAPTAIN OF THE GORGON
NOT A DAY MAN,
BUT A MAN FOR ALL TIME.

THE CREWS OF THE TELEGRAPH FLEET.
THE "UNION JACKS" THAT WILL
COMMAND RESPECT WHEREVER THEY FLOAT."

Also, a design representing Science and the laying of the Cable, with the quotation from Shakspeare, "I'll put a girdle round about the earth in forty minutes."

At the Webster House paper lanterns and a banner with the motto:

"THE NIAGARA, A FLOATING MONUMENT TO GEORGE STEERS."

The New York Hotel had a transparency on which were

designed an Englishman pulling by the horns a bull, to whose tail was attached the Telegraph Cable; an Irishman on a rock in the centre, and an American at the right hand side paying out the Cable. The Englishman says: "I'm afraid it's all up with us, Mr. Bull—the Yankee's got us." The complaisant Hibernian to the American says: "Pull away, Jonathan; ah, be jabers! if the tail houlds ye'll fetch 'em." The American replies: "Keep cool, Pat, and I reckon we'll annex you both." Under this was the motto:

"EXTENDING THE AREA OF FREEDOM."

At 753 Broadway there were six transparencies in medallion form, representing John Bull and Brother Jonathan at fisticuffs in 1776. Jonathan on a slow nag, with the inscription, "The Past." Britannia—"I'm monarch of all I survey." In 1858 Brother Jonathan and John Bull are embracing each other ardently, and are bound round with something supposed to be a telegraphic wire, but which looks very much like crinoline. Then comes Mercury striding the waters with news, and underneath, the inscription, "The Present." And last, Old Father Time, who says, "My dominion is invaded."

At No. 754 a transparency representing Franklin flying a kite, and the Niagara laying the Cable—with

"Franklin,<br>Morse, Maury,<br>Field, Hudson,<br>Everett, Steers. } *Committee on Lightning.*"

At No. 757, banner, the Eagle—the Victory of War. The Lion and Unicorn—the Victory of Peace.

At No. 788 Broadway, the following inscriptions were displayed:

"THE NIAGARA,
CAPTAIN HUDSON.
VICTORIA."

"THE AGAMEMNON,
CAPTAIN PREEDY.
BUCHANAN."

"Full many a gem of purest ray serene,
The dark unfathomed caves of ocean bear.

THE CONQUEST.

*Eripuit Coelum Fulmens.*

FRANKLIN,

TERRA,
MARE,

MORSE.
FIELD."

The Spingler Institute (Union Square) had the American flag suspended over a banner, with the following inscription:

"THE SPINGLER INSTITUTE.
THE DAUGHTERS OF AMERICA
SEND GREETING
TO
THE DAUGHTERS OF ENGLAND."

Such, to be brief, was the character of the decorations which private enterprise displayed on Broadway. During the day they had countless admirers, but their full effect was unknown until after nightfall, and the gas-jets, lanterns, candles and transparencies were lit up, flooding the street with a blaze of light that was overwhelming.

THE SERVICES AT TRINITY CHURCH.

But we must recur to the programme:

Pursuant to previous arrangement and notice, a grand *Te Deum*, with other services, was celebrated in Trinity Church, at ten o'clock, in the morning.

The external decorations of the edifice were few, but highly appropriate. Rev. Dr. Ogilby thought it would be proper to signalize the occasion by the display of bunting from the noble spire. Accordingly, he addressed a note to Commodore Kearney, of the Brooklyn Navy Yard, asking him if he could furnish a few colors. The response was the following kind note:

NAVY YARD, NEW YORK, August 27, 1858.

*To the Rev. Frederick Ogilby, D. D., Assistant Rector of Trinity Church, New York City:*

DEAR SIR—As Commandant of the Navy Yard, New York, it would afford me great pleasure to furnish the flags you asked to be loaned, for the celebration of the laying of the Atlantic Telegraph Cable. The flags of this yard have already been disposed of for the same purpose, but the "NIAGARA'S FLAGS" having just been landed, I am happy to have it in my power to comply with the Rector's wish. And I thank you, sir, very much for the invitation to the officers and myself to be present at a service so distinguished at Old Trinity.

I am, very respectfully,
Your obedient servant,
LAWRENCE KEARNEY.

So the Niagara's colors were furnished—and beautiful flags they were. The St. George's Society, St. Andrew's Society, Mr. Dale, of the Liverpool Propeller line, and several other gentlemen, also contributed flags. In accordance with the recommendations of the architects of the church, to whom the suggestion of the general plan of the decorations was committed, the American and English flags were displayed, side by side, from the two upper spire lights, some two hundred and thirty feet from the sidewalk. Above the battlements of the tower, over a hundred and twenty feet from the street level, flags of other nations were displayed in groups, on each side, three together. On the north side were the French, Spanish and Austrian; on the south side, the Russian, Swedish and Norwegian; on the east side, looking down Wall street, the Dutch, Neapolitan and Potuguese; on the west side, overhanging the body of the church, the Brazilian, Mexican and Chilian. Their appearance was so novel that it drew an immense crowd of people to gaze at the spectacle. In the porch of the church were festooned two superb flags—one, the standard of the St. George's Society, and the other, the ensign of the Niagara. These flags were suspended from the centre of the porch roof, and were gathered up at either side. Beneath them the grand procession of the clergy and Corporation entered the church.

At nine o'clock the flags were displayed, and at half-past nine, precisely, Mr. James E. Ayliffe, the bellringer of the church, began to ring the chimes, according to the following programme:

1. Changes on Eight Bells.
2. Hail Columbia.
3. God Save the Queen.
4. Yankee Doodle.
5. Evening Hymn.
6. Mariner's Hymn.
7. The Verona.
8. Vesper Hymn.
9. London (new).
10. Bedford.
11. Old Hundred.
12. Evening Bells.
13. Yankee Doodle.
14. God Save the Queen.
15. Hail Columbia.
16. Changes on Eight Bells.

The flags and chimes soon drew together an immense throng of people, who pressed hard for admission. But, by order of the Rector, the gates were kept closed, and no person, except the Rev. the clergy who were to participate in the services, was admitted without a ticket, before half-past nine, at which time the doors were thrown open to the public. A force of fifteen picked patrolmen under Sergeant Croft, of the Mayor's Office, served to guard the gates, while a strong force of policemen kept the throng on Broadway in order and opened a way for the invited guests.

The entire centre aisle of the church was reserved for

the accommodation of the invited guests, the most notable of whom were as follows:

The Mayor and Corporation of New York.
The Mayors and Corporations of Brooklyn, Jersey City and Hoboken.
General Scott.
Lord Napier's Suite.
H. B. M. Consuls of New York and Charleston.
H. B. M. Vice-Consul at this Port.
Federal Officers.
Washington Irving.
Lords Ribblesdale, Cavendish and Grosvenor.
Com. Kearney, U. S. N., and Officers of the Navy Yard.
Montreal Military Officers.
Judges of the State Courts.
Officers of the Atlantic Telegraph Co.
Capt. Dayman and Officers of the Royal Navy.
Capt. Hudson and Officers of the Niagara.

Mr. Field was not present, having pressing previous engagements which he could not lay aside. A number of distinguished gentlemen, of both nations, were absent for similar reasons. The guests who attended, generally arrived before half-past nine, the hour for the opening of the gates. The Corporate authorities of New York and Brooklyn, however, entered in procession, with their staves of office, after the rest of the congregation had been seated.

The gates were thrown open at half-past nine o'clock precisely, when an immense rush was made by the crowd to enter. It was with the utmost difficulty that the police could enforce the decorum due to the solemnity of the building and the occasion. Every unreserved place in the church was almost instantly filled by the people, and not an hundredth part of those in the street found accommodation. A dense throng remained outside during the whole service.

As the City Corporation marched in, the bells chimed the national airs.

The clergy assembled in the vestry-rooms of the church about a quarter past nine. So far as could be ascertained, the following is a correct list:

The Rt. Reverend
The Bishop (Doane) of New Jersey.

NEW YORK.

Rev. H. Anthon, D. D.
" Edward Anthon.
" J. Graeff Barton.
" Alfred B. Beach, D. D.
" G. T. Bedell, D. D.
" Wm. Berrian D. D.
" James Bonnar.
" J. H. H. Brown.
" N. W. Camp, D. D.
" P. S. Chauncey, D. D.
" Caleb Clapp.
" A. V. Clarkson.
" F. D. Harriman.
" S. Cooke, D. D.
" N. E. Cornwall.
" Richard Cox.
" C. F. Cruse, D. D.
" S. D. Denison.
" R. G. Dickson.
" Morgan Dix.
" G. B. Draper.
" C. R. Duffie.
" H. Dyer, D. D.
" Lot Jones.
" F. E. Lawrence.
" A. S. Leonard.
" Chas. S. Little.
" Wm. S. Ludlum.

Rev. T. A. Eaton.
" W. E. Eigenbrodt, D. D.
" J. J. Elmendorf, D. D.
" A. Fitch.
" E. O. Flagg.
" Wm. J. Frost.
" T. Gallaudet.
" George J. Geer.
" John Grigg.
" C. W. Hackley, D. D.
" Richard C. Hall.
" A. Bloomer Hart.
" F. L. Hawks, D. D., LL.D.
" E. Y. Higbee, D. D.
" J. H. Hobart, D. D.
" J. H. Hopkins, Jr.
" G. H. Houghton.
" R. S. Howland.
" Ralph Hoyt.
" Hiram Jelliff.
" S. R. Johnson, D. D.
" S. Seabury, D. D.
" F. W. Smith.
" J. A. Spencer, D. D.
" Alpheus Spor.
" F. W. Taylor, U. S. N.
" T. H. Taylor, D. D.
" B. F. Taylor.

Rev. Milo Mahan, D. D.
" E. C. Marshall.
" J. McVickar, D. D.
" James Millett.
" H. E. Montgomery.
" John Morgan.
" W. F. Morgan, D. D.
" W. Morris, LL. D.
" W. A. Muhlenberg, D. D.
" Geo. W. Nichols.
" F. Ogilby, D. D.
" B. C. C. Parker.
" George C. Pennell.
" T. McC. Peters.
" C. E. Phelps.
" J. H. Price, D. D.

Rev. Thomas Towell.
" S. H. Turner, D. D.
" Isaac H. Tuttle.
" Robert Travis, Jr.
" R. B. Van Kleeck, D. D.
" Antoine Verren.
" J. M. Waite.
" Wm. Walton, D. D.
" Henry D. Ward.
" F. S. Wiley.
" M. Evans Willing.
" B. Wilson, D.D., LL. D.
" Calvin Wolcott.
" J. Freeman Young.
" J. S. Purdy.
" Wm. Richmond.

Rev. Wm. G. Farrington.

RIVER TOWNS.

Rev. J. S. Spencer.

Rev. F. Babbit.

BROOKLYN.

Rev. James S. Barnes.
" Lemuel Burge.
" Eli H. Canfield, D. D.
" J. F. D. Cornell.
" Isaac F. Fox.
" B. C. Cutler, D. D.
" Edward DeZeng.
" Jacob W. Diller.
" T. S. Drowne.
" Edmund Embury.
" Jared B. Flagg.
" Chas. H. Gardiner.
" Henry T. Gregory.
" Thomas T. Guion.

Rev. D. V. M. Johnson.
" Evan M. Johnson.
" Isaac P. Labagh.
" W. H. Lewis, D. D.
" J. A. Paddock.
" Francis Peck.
" O. Perinchief.
" Chas. S. Putnam.
" E. F. Remington.
" W. W. Seymour.
" Henry A. Spafard.
" F. Tripp.
" Robt. J. Walker.
" E. Wheeler.

WILLIAMSBURGH.

| | |
|---|---|
| Rev. Alvah Guion. | Rev. A. H. Partridge. |
| " Samuel M. Haskins. | " Merritt H. Wellman. |

NEW JERSEY.

| | |
|---|---|
| Rev. Charles Arey. | Rev. Charles H. Canfield. |
| " James J. Bowden. | " O. Harriman, Jr. |
| " Vandervoort Bruce. | " N. Sayre Harris. |

Rev. Wm. R. Earle.

STATEN ISLAND.

| | |
|---|---|
| Rev. J. C. Eccleston, D. D. | Rev. Jesse Pound. |
| " Theodore Irving. | " H. L. E. Pratt. |
| " Pierre P. Irving. | " G. Winslow, D. D. |
| " Frederick Oertel. | " Mr. Shackelford. |
| " Mr. Gordon. | " Mr. Cady. |
| " Mr. Croes. | " E. M. Peck. |

Rev. C. R. Smith.

TEXAS.

Rev. Benj. Eaton.

The ministers brought their surplices, and robed in the vestry-room.

About the same hour the organists and singers arrived and repaired to the organ loft. The choir was carefully selected from those of the principal Episcopal churches in the city. The following is a list of the leading vocalists:

TREBLES.

Miss Gellie, of Calvary Church.
Miss Goodwin, of Trinity Chapel.
Mrs. Hutchings, of Trinity Church.
Mrs. Walter, of Trinity Chapel.
Mrs. Bourne, of the Church of Holy Apostles.

ALTOS.

Mr. Granville, of Trinity Church.
Miss Mary Gellie, of Calvary Church.
Miss Hodges, Director of Music, Christ Church, Norwich, Conn.
Miss Robjohn, St. Thomas' Church.

TENORS.

Mr. Deane, of Trinity Church.
Mr. Peck, of Trinity Church.
Hr. Hurley, of Trinity Church.
Mr. Good, of Trinity Chapel.
Mr. Gledhill, Jr., of Trinity Chapel.

BASSES.

Mr. Browne, of Trinity Church.
Dr. Guilmette, of Dr. McAuley's Church, Fifth avenue.
Mr. Cooke, of Trinity Chapel.
Mr. Safford, of Trinity Church.

BOYS.

Gamble, (2) of Trinity Chapel.
James Brown, of Trinity Church.
H. E. Brown, of Trinity Church.
Sidney Terhune, of Trinity Church.

These were well sustained by an accomplished chorus.

Edward Hodges, *Mus. Doc.*, Director of Music in the Parish, presided at the Organ, assisted by Mr. Wm. Walter, organist of Trinity Chapel, and Mr. J. F. Huntington, organist of St. Thomas' Church.

The internal decorations of the church were of the most novel and magnificent description. Across the chancel, between the jambs of the grand arch, was thrown a temporary screen, consisting of three pointed gothic arches, twenty-three feet high, sustaining a horizontal panel four feet high divided into three compartments. Above this sprung a lofty central gothic arched panel eighteen feet high, surmounted by a floriated cross seven feet high. The whole screen was forty-eight feet high to the top of the cross. The columns supporting the arches, the arches themselves, and all the lines of construction were heavily covered with evergreens; white and red cedar, white pine, larch, hemlock, spruce, fir, box, holly and several other evergreens of various hues being artistically combined, completely hiding any trace of the wooden frame. The whole surface of the structure was covered at short intervals with magnificent bunches of the most exquisite hot-house and garden flowers, of rich bright colors and delicious perfume. The varieties of flowers brought into requisition in the construction of the wonderful screen were almost fabulous in numbers. With lavish profusion the rarest and most exquisite specimens of the floral kingdom were displayed.

Bunches of these flowers were arranged artistically on the columns that supported the arches, gracefully swelling the body of the pillars, and almost obscuring the green ground. The three arches were also profusely covered with flowers, which were arranged in figures of light gothic tracery in the openings. The horizontal panel over the arches was of fine white linen, divided into three compartments by broad boarders of evergreens, tipped with roses, lilies, gladiolas, dahlias and other gorgeous flowers. Among the dahlias was an entire new variety, now introduced for the first time, and named, in honor of the occasion, the "Cable Dahlia." It is a large double flower, five inches in diameter, of a yellow ground color, ring-streaked, striped, speckled, and spotted with various hues of crimson and scarlet.

The lofty central arched panel was four feet wide, of

white linen like the horizontal panel, and similarly bordered with evergreens and flowers. In this panel appeared the inscription,

"GLORY BE TO GOD ON HIGH,"

in letters of flowers, the words "Glory be to" on the right base of the arch, the word "God" around the point of the arch, and the "on high" on the left base. Each word, excepting the title of the Deity, was in flowers of one color. The words "glory be to" and "on high" were of scarlet sword lily and dark roses; the word "God" was in large capitals, of white roses set upon red roses. In the right compartment of the horizontal panel were the words,

"AND ON EARTH,"

in gothic capitals, composed of roses, of various hues, blue hydrangeas and verbenas. In the central compartment were the words,

"PEACE, GOOD-WILL,"

in gothic capitals of scarlet zinnias, white phlox and Chinese asters. In the left compartment were the words, in Gothic capitals,

"TOWARDS MEN,"

the flowers being similar to those in the first compartment. In the space inclosed by the central arched panel was a quatrefoil, aptly symbolizing the four quarters of the world, whose bringing together was the occasion of the services. The quatrefoil was covered with evergreens, and relieved by an immense variety of flowers, prominent among which were dahlias, of many hues, roses, of several varieties, salvia splendens, and the new and superb (blue) delphinum formosum. In the centre of the quatrefoil was a five-pointed star, intended to symbolize the five races of men.

It was composed of evergreens, and ornamented with flowers. The centre of the star was composed of scarlet dahlias; next was a ring of white phlox, and the points were of beautiful blue hydrangea. The extreme tips of the points were rose-colored dahlias.

The cross on the top of the lofty arched panel was the crowning work of the whole structure. It rested upon a cushion of scarlet lilies. The whole body of the cross was chastely formed of pure white roses, phloxes, lilies etc. In the centre, where the arms crossed the upright, was a circle of scarlet dahlias; and the points of the head and arms were of scarlet salvia. The top of the cross was fifty feet from the floor of the church, and forty-eight feet from the foot of the screen. The effect of the whole structure was almost indescribable, especially when lighted up by the morning sun streaming down from the clerestory windows, and standing out against the subdued and solemn colors of the great chancel window, which was seen here and there through the airy tracery of the screen.

The bronze eagle lectern stood in the centre arch of the screen. In the eagle's mouth was a ring of the genuine Atlantic Cable, bound with gold. A bouquet of rose, shamrock and thistle, bound with ribbons of red, white and blue, was attached to the ring.

The font was covered with an octagonal canopy, ten feet high. The canopy formed a pyramidal bouquet, composed of the rarest floral gems, the principal colors being blue, scarlet, white and crimson, gracefully interspersed with green leaves.

The pulpit was ornamented with a delicate tracery of flowers about the banisters. Borders of green, with wreaths of flowers running up the angles, were supplied to all the panels, and on every angle was a bouquet as a finil. The *tout ensemble* of these floral decorations was unprecedentedly striking, and the execution of the design reflected the highest credit upon the taste of the artists who were engaged upon it.

The whole neighborhood of New York was scoured for the flowers, and every greenhouse or garden of any prominence within twenty miles furnished its tribute to the grand design. The perfume of all these exquisite flowers was almost overpowering, and filled every corner of the church with the most delicious aroma.

On the second and third pillars, on the right side of the nave, looking from the chancel, large English flags were festooned over the seats occupied by the English diplomatic representatives and officers. On the two corresponding pillars, on the opposite side of the nave, over the seats occupied by the officers of the Niagara and the Directors of the Telegraph Company, were suspended two American flags.

At ten o'clock the clergy, in their surplices, formed in the vestry-room, and, the bells chiming "Old Hundred," they marched, two and two, out of the north door of the church, through the church-yard along the northern wall to the grand doorway. The sexton, with his mace, preceded the clergy, bareheaded; next came the deacons, two by two, next the priests, two and two, and the Rt. Rev. the Bishop brought up the rear.

As soon as the procession had reached the grand entrance, the organ burst forth in a joyous voluntary. The congregation rose and turned towards the middle aisle, while the procession marched up the aisle until the deacons reached the steps of the chancel. The clergy then halted, and faced inwards, open order, while the sexton, bearing his mace aloft, slowly walked down between the lines to meet the Rt. Rev. the Bishop at the door. Taking his place before the prelates, he countermarched to the chancel, the priests following in regular order and closing the procession after the Bishop. The Prelate entered his chair and the officiating clergy the regular stalls, while the great mass of the clerical gentlemen present occupied temporary stalls back of the screen. When all had taken their places the music suddenly ceased, and all—clergy and congregation—knelt in silent prayer.

After a pause of an instant, the organ and full choir burst out in the introductory anthem :

"The Lord is in his holy temple; let all the earth keep silence before him."—*Hab.* ii. 20.

This was given in full chorus, to music composed by Dr. Hodges for the service at the consecration of Trinity Church in the year 1846. The other opening sentences of Morning Prayer,

"From the rising of the sun even unto the going down of the same, my Name shall be great among the Gentiles; and in every place incense shall be offered unto my Name, and a pure offering: for my Name shall be great among the heathen, saith the Lord of hosts."—*Mal.* i. 11.

"Let the words of my mouth, and the meditation of my heart, be alway acceptable in thy sight, O Lord, my strength and my redeemer."—*Psalm* xix. 14, 15.

were then said by Rev. Dr. Creighton, Rector of Christ Church, Tarrytown, Rector of St. Mary's Church, Beechwood, and Chairman of the General Convention. Dr.

Creighton also said the exhortation, and led in the General Confession. The choir, with organ accompaniment, chanted the

*Venite Exultemus Domino.*

"O come, let us sing unto the Lord; let us heartily rejoice in the strength of our salvation.

Let us come before his presence with thanksgiving; and show ourselves glad in him with psalms.

For the Lord is a great God; and a great King above all gods.

In his hand are all the corners of the earth; and the strength of the hills is his also.

The sea is his and he made it; and his hands prepared the dry land.

O come, let us worship and fall down; and kneel before the Lord our Maker.

For he is the Lord our God; and we are the people of his pasture, and the sheep of his hand.

O worship the Lord in the beauty of holiness; let the whole earth stand in awe of him.

For he cometh, for he cometh to judge the world, and the people with his truth.

Glory be to the Father, and to the Son, and to the Holy Ghost.

As it was in the beginning, is now, and ever shall be, world without end.—*Amen.*"

The music for the *Venite* was an octave chant by Jones, much admired by Haydn when he heard it sung by three thousand children in St. Paul's Cathedral, London. It was a grand chant, and most correctly rendered, the clergy and congregation standing.

The Tenth selection of Psalms from the Psalter was next said by the Rev. Dr. Creighton, the congregation standing and responding. The psalms of this selection were peculiarly appropriate to the occasion, and the opening and closing psalms were as follows :

FROM PSALM xcvi.—*Cantate Domino.*

"O sing unto the Lord a new song; sing unto the Lord, all the whole earth.

Sing unto the Lord, and praise his name; be telling of his salvation from day to day.

Declare his honor unto the heathen; and his wonders unto all people.

For the Lord is great, and cannot worthily be praised; he is more to be feared than all gods."

PSALM cl.—*Laudate Dominum.*

"O Praise God in his holiness: praise him in the firmament of his power.

Praise him in his noble acts: praise him according to his excellent greatness.

Praise him in the sound of the trumpet: praise him upon the lute and harp.

Praise him in the cymbals and dances: praise him upon the strings and pipe.

Praise him upon the well-tuned cymbals: praise him upon the loud cymbals.

Let every thing that hath breath praise the Lord."

The Laudate Dominum was magnificently chanted by the choir, with powerful organ accompaniment, and immediately upon its conclusion, the choir burst out into the

*Gloria in Excelsis.*

"Glory be to God on high, and on earth peace, good will towards men. We praise thee, we bless thee, we worship thee, we glorify thee, we give thanks to thee for thy great glory, O Lord God, heavenly King, God the Father Almighty.

O Lord, the only begotten Son Jesus Christ; O Lord God, Lamb of God, Son of the Father, that takest away the sins of the world, have mercy upon us. Thou that takest away the sins of the world, have mercy upon us. Thou that takest away the sins of the world, receive our prayer. Thou that sittest at the right hand of God the Father, have mercy upon us.

For thou only art holy; thou only art the Lord; thou only, O Christ with the Holy Ghost, art most high in the glory of God the Father. --*Amen.*"

After the *Gloria*, the congregation were seated, and the Rev. Dr. Hawks, Rector of Calvary Church, read the First Lesson from ISAIAH xl. 3. The most striking verses of the chapter were the following:

2. "When thou passest through the waters I will be with thee; and through the rivers, they shall not overflow thee; when thou walkest through the fire, thou shalt not be burned; neither shall the flame kindle upon thee."

16. "Thus saith the Lord, which maketh a way in the sea, and a path in the mighty waters;

17. Which bringeth forth the chariot and the horse, the army and the power; they shall lie down together, they shall not rise; they are extinct, they are quenched as tow."

19. "Behold, I will do a new thing; now it shall spring forth; shall ye not know it? I will even make a way in the wilderness, and rivers in the desert."

After the lesson came the great feature of the service, that sacred hymn,

*Te Deum Laudamus.*

"We praise thee, O God; we acknowledge thee to be the Lord.
All the earth doth worship thee, the Father everlasting.
To thee all Angels cry aloud; the Heavens, and all the Powers therein.
To thee Cherubim and Seraphim continually do cry,
Holy, Holy, Holy, Lord God of Sabbaoth;
Heaven and earth are full of the Majesty of thy glory.
The glorious company of the Apostles praise thee.
The goodly fellowship of the Prophets praise thee.
The noble army of Martyrs praise thee.
The holy Church throughout all the world doth acknowledge thee
The Father of an infinite Majesty;

Thine adorable, true, and only Son;
Also the Holy Ghost, the Comforter.
Thou art the King of Glory, O Christ.
Thou art the everlasting Son of the Father.
When thou tookest upon thee to deliver man, thou didst humble thyself to be born of a Virgin.
When thou hadst overcome the sharpness of death, thou didst open the kingdom of Heaven to all believers.
Thou sittest at the right hand of God, in the glory of the Father.
We believe that thou shalt come to be our Judge.
We therefore pray thee help thy servants, whom thou hast redeemed with thy precious blood.
Make them to be numbered with thy saints, in glory everlasting.
O Lord, save thy people, and bless thine heritage;
Govern them, and lift them up forever,
Day by day we magnify thee;
And we worship thy name ever, world without end.
Vouchsafe, O Lord, to keep us this day without sin.
O Lord, have mercy upon us, have mercy upon us.
O Lord, let thy mercy be upon us, as our trust is in thee.
O Lord, in thee have I trusted, let me never be confounded.

This *Te Deum* was a "verse service," in the key of D, commonly known as the "New York Service," and was composed by Dr. Hodges in 1840. It was a varied composition, interspersed with solo and duet passages, and written in the English style of cathedral music, though rather more florid than most of the English service. It opened with a full chorus in plain counterpoint, in a majestic and dignified style. This was succeeded by a fugue passage at the words, "The glorious company of the Apostles praise Thee," in which the subject was admirably carried through all the voices, while the immediately succeeding verses were announced without repetition of words, arriving at a grand climax at the verse "Thou art the King of Glory, O Christ." Here the magnificent outburst of harmony, from the full organ and choir, was exceedingly impressive.

The rest of the *Te Deum* was an alternation of solos, duets and choruses.

Rev. Dr. Bedell, of the Church of the Ascension, then read the Second Lesson from Revelations iv., the most appropriate verses being as follows:

10. "The four and twenty elders fall down before Him that sat on the throne, and worship Him that liveth for ever and ever, and cast their crowns before the throne, saying,

11. Thou art worthy, O Lord, to receive glory and power; for thou hast created all things, and for thy pleasure they are and were created."

After the Lesson the congregation rose, and the choir sung the

*Benedictus.* St. Luke i., 68.

"Blessed be the Lord God of Israel; for he hath visited and redeemed his people;

And hath raised up a mighty salvation for us, in the house of his servant David;

As he spake by the mouth of his holy prophets, which have been since the world begun;

That we should be saved from our enemies, and from the hand of all that hate us."

The music was part of the same service in "D" to which the Te Deum belonged, and was much in the same style, with perhaps greater play of fancy for the organ accompaniment.

The Rev. Dr. Creighton then led in the Apostle's Creed, after which all kneeled down, and the Rev. the Rector of the Parish, the venerable Dr. Berrian, said the balance of

Morning Prayer to the minor benediction. Just before the General Thanksgiving, he said the following

SPECIAL PRAYER FOR A CABLE SERVICE.

*(Allowed by the Bishop.)*

"O God, whose never-failing Providence ordereth all things, both in heaven and earth; we, thy humble servants, bow before thee, owning that from thee all strength, all wisdom, power and might do come. We praise thee for thy goodness and wonderful works to the children of men, and acknowledge thy gracious hand in all that we accomplish upon earth. Especially this day do we recognize thy goodness and mercy in the wonderful work for which we now bless and magnify thy glorious name. Thou, who alone spreadest out the heavens and rulest the raging of the sea, didst in thy mercy guide thy servants through the perils of the great deep, and enable them to lay in the mighty waters that bond which now unites distant nations. Grant, O Lord, that those who are so wonderfully joined together may never be put asunder by enmity or strife, by prejudice or passion. May it be instrumental in bearing only the messages of peace, extending the glad tidings of salvation, the gospel of thy dear Son, and hastening the day when from every corner of the earth shall rise that blessed song, "Peace on earth, good-will towards men." And to thy great name shall be ascribed all honor and praise, through Jesus Christ, our Lord." *Amen.*

After the General Thanksgiving, the Rev. Rector said the following Special Thanksgiving, at the request of Capt. Hudson and the officers of the Niagara:

FOR A SAFE RETURN FROM SEA.

"Most gracious Lord, whose mercy is over all thy works; we praise thy holy name that thou hast been pleased to conduct in safety, through the perils of the great deep, these servants, who now desire to return their thanks unto thee' in thy holy Church. May they be duly sensible of thy merciful providence towards them, and ever express their thankfulness by a holy trust in thee, and obedience to thy laws; through Jesus Christ our Lord." *Amen.*

Rev. Dr. Ogilby announced that the Lord Bishop of Montreal, who was expected to represent the English Church, was absent unavoidably, on account of engagements made prior to the announcement of this service. The Rt. Rev. the Provisional Bishop was absent, for a similar reason.

Rev. Dr. Ogilby then announced the Anthem, Psalm cxxxiii. of the Psalter:

PSALM cxxxiii. *Ecce, quam bonum!*

"Behold, how good and joyful a thing it is, brethren, to dwell together in unity!

2. It is like the precious ointment upon the head, that ran down unto the beard, even unto Aaron's beard, and went down to the skirts of his clothing.

3. Like as the dew of Hermon, which fell upon the hill of Sion.

4. For there the Lord promised his blessing, and life for evermore."

This was sung to an anthem composed by Dr. John Clark Whitfield. It was a pleasing but an unpretending work, selected upon this occasion on account of the peculiar appropriateness of the words, "Behold, how good and joyful a thing it is, brethren, to dwell together in unity."

Rev. Dr. Ogilby then announced that an address would be delivered by the Right Rev. the Bishop of New Jersey. The distinguished prelate in his robes took a position just under the central arch of the screen, in the rear of the lectern, and spoke, with great deliberation and emphasis, as follows:

"Glory be to God on high, and on earth peace, good-will towards men."

This was the message of the angels to the shepherds on the plain of Bethlehem, when the incarnate Saviour of the world was cradled in the manger. This was the message of the angels, by the Atlantic Telegraph, to their western sons; and this shall be the Anglo-American message to the ends of the whole earth, "Glory be to God on high, and on earth peace, good-will towards men." Was ever utterance so fit? Was ever utterance so startling? Was ever utterance so solemn, so sublime? Flashing out from the burning realm of Christian hearts in Ireland; flashing along through the caverns of the sea; flashing along among the buried treasures of the deep; flashing along through the lair of old Leviathan; flashing along among the remains of them who perished in the flood; flashing up among the primeval forests of Newfoundland, and flashing out from there throughout the world! A consecrated lightning, consecrating the very ocean through which it traverses; consecrating this glorious, blessed day; consecrating anew that time-honored flag of England—the banner of a thousand fights; consecrating the stars that glitter on the flag of freedom, which, in less than a century, has won for this nation its place among the ancient empires of the world, and which, wherever the rights of men are to be asserted, forever floats and blazes in the van. Consecrating, shall I not say, beloved friends, consecrating anew our hearts to the love of man and to the glory of the living God. It is recorded of the father of Hannibal, that he took his son, almost an infant, to his heathen altar, to swear eternal hatred against Rome. Shall we not come up here to-day—have we not come up here to-day, to renew before this holy altar our vows of love and peace? Shall we not here renew the vows of our baptism, that so far as in us lies we will live peaceably together; that, so far as in us lies, we will promote that which makes peace, and quietness, and love among all men; that, so far as in us lies, each in his several place, by services, by suffering, by death, if God please, we will do what lies in us to bear out to all the world lying in darkness, wickedness, and sin, the peace and love of the Gospel of our Lord and Saviour Jesus Christ. It seems to me, if I may speak it without irreverence, that oneness is the great idea of God—oneness is the great idea of God. The unity of God is the great truth. "There are three that bear record of heaven—the Father, the Son, and the Holy Ghost—and these three are one." And again, "I and the Father are one." And in the beseeching prayer, as He was about to enter into the garden of agony, that "they might be one, even as we are." "I in Thee, and Thou in me, that they also may be one in us." St. Paul instructs us that there is "one body and one spirit, one God and Father of all—one love, one faith, one baptism;" and then, and then only will the mediatorial glory be accomplished, when there shall be one fold and one shepherd, Jesus Christ our Lord. Now, it seems to me that, among the thousand thoughts that crowd upon the mind in the contemplation of the great subject of this day's assembling, the tendency to oneness is the chief. It seems to me that, in a sort, the edict of Babel is reversed—that the dispersion of the nations is to be undone, in God's time and in God's way, by bringing them together as one in him; and I might almost venture to say, that we have in prospect, as it were, the renewal and repetition of the Pentecostal wonder, when all the nations of the world heard in their own tongue the wonderful works of God; when man shall speak to man, from the one end of the world to the other, of the gospel of the Saviour and of the glory of the Lamb. Oneness—its tendency to oneness is the leading thought, in my mind, as the result of the great event which we celebrate this day. Beloved friends, I have come among you to-day, traveling through the night to be with you, from the field of my own labors and from the care of my two hundred children, that with my brethren and companions I might worship in this holy and beautiful house, and with them, and with you all, and with all England, and with all Europe, and with the islands of the sea, rejoice in the consummation of this great work. Beautifully and well did this venerable Corporation seek for itself a place in the rejoicings of this day. Trustees they are from venerable hands in that dear mother land, now gathered to the grave ; trustees they are for carrying out their views and purposes, and, great and glorious as are the good works which they have done, they have done none greater or more glorious than in lending the consecration of this house, the consecration of that altar and these prayers to the Atlantic Telegraph. I came from New Jersey, and I have brought something of New Jersey with me. I have brought the youngest child of the telegraph. This (exhibiting a piece of wire) is the germ which has grown to what it now has become—so great and glorious. So far as I know and believe, this is a part of the telegraph wire set up at the Speedwell Iron Works, in Morristown, New Jersey, more than twenty years ago, under the direction of Professor Morse, known to all the world, and Mr. Alfred Gray, his associate and co-laborer. It was set up for a length of three miles, and it served to transmit intelligible signals in telegraphic language. This has nothing to do, by comparison or contrast, with what we celebrate to-day. The acorn is not the oak—the germ is not the tree—the infant is not the man. We rejoice now in the full stature of the man—in the tall glory of the palm—in the shading verdure of the monarch oak ; and we ascribe, under God, the practical application of that, which was felt after so long, as is the case in every great invention—we ascribe the practical result, under God, to one Cyrus, to his energy and devotion, under God ; it is by noble souls in both hemispheres that the chain which binds together the two continents was successfully laid. Space is, as it were, annihilated, and time more than annihilated. In a sense there is no more sea. As I stand here I feel that I can lay my hand upon the tomb of Chaucer. We can go with holy George Herbert to hear the "Angel's Music" from the bells of Salisbury. We may breathe the air made fragrant by the dying breath of Latimer and Ridley; our children can unite with England's children in the song of faith; and the men of the West may stand up with those of the East, when they say "I believe in God the Father Almighty, maker of heaven and earth." We have all read of that beautiful ceremony which was once celebrated—the wedding of the Adriatic Venice and the fleet of gondolas has made a picture on every heart of childhood. Venice is no longer among the nations. The glory of the Adriatic has departed. But now follows another wedding. The day breaks upon the rugged shores of Newfoundland. A little company is landing from a boat. They form a line. They bear in their hands, and touch it as a sacred thing, a small wire, and they proceed with solemn step and slow to the place appointed to deposit it. Cyrus at the head, they form a procession in comparison with which the heroes of antiquity must look to their laurels. Carefully they proceed, charged, as they feel, not only with the destiny of nations, but with the Church of the living God, and repose it in the proper place—a gallant sailor, captain in our navy, surrounded by officers of our sister navy and by the sailors of both fleets—an act inimitable in beauty, and a testimony that God was with them in truth—they pour out their hearts in prayer to God, thanking him for his mercies and asking him for his blessing; and then with a shout they awake the virgin echoes in rejoicings for the consummation of that great event which has made two into one; which has wedded England and America, and brought them together—together for civil freedom; together for the progress of knowledge; together for the happiness of home; together for the extension of the Gospel; together for the edification of the Church; together for the salvation of the world; together to bring about that glorious time when angels shall again descend from heaven, and the whole redeemed world, with all the company of angels and archangels, shall lift up once more that glorious hymn —"Glory to God in the highest; on earth peace, good-will toward men." England and America have been wedded by the Atlantic ring—a ring of love—a ring of peace—shall I not say a ring of God? Shall I not say it? Will not every heart respond—amen? "Those whom God hath joined together, let no man put asunder."

When the Bishop had concluded, Rev. Dr. Berrian gave out the One Hundredth Psalm:

"With one consent let all the earth
To God their cheerful voices raise;
Glad homage pay with joyful mirth,
And sing before him songs of praise—

"Convinced that he is God alone,
From whom both we and all proceed;
We, whom he chooses for his own,
The flock that he vouchsafes to feed.

"Oh, enter, then, his temple gate,
Thence to his courts devoutly press;
And still your grateful hymns repeat,
And still his Name with praises bless.

"Praise God, from whom all blessings flow;
Praise him all creatures here below,
Praise him above, ye heavenly host,
Praise Father, Son, and Holy Ghost."

This was sung to the old One Hundredth Psalm tune. Dr. Hodges' accompaniment was played with the whole force of the organ, and the choir sang in full chorus for the first two lines, by which time every voice within the walls of the great church joined in. At the same instant the bell-ringer struck up the Psalm upon the chimes, and with the power of the magnificent organ of the church, the bells, and a chorus of three thousand five hundred voices—many of them male—Old Hundred was given with a hearty majesty that has rarely been accorded to it anywhere. When the last note had died away, the congregation kneeled and the Rt. Rev. the Bishop pronounced the major benediction.

Dr. Hodges then played, as a concluding voluntary, Handel's anthem, the words "Zadock, the Priest, and Nathan, the Prophet."

Rev. Dr. Ogilby came forward and said that Mayor Tiemann had put in his hands a message received from Savannah, which so singularly accorded with the address just delivered, that it would seem as though there was a telegraph at work between them. He would read it:

SAVANNAH, September 1, 1858.

TO MAYOR TIEMANN:

Savannah joins her sister city in a chorus of joy and gratitude for the blessings that have joined what nature seemed to have eternally sundered. The Anglo-Saxon race has made the lightning of heaven the swift messenger of peace. Our nation is clasped in the embrace of friendship by our former enemy. In all this we see the power of Providence guiding the nations in the way of peace, and the two great branches of our race instruments in His hand.

THOMAS TURNER,
*Mayor of Savannah.*

The services being now over, the clergy again formed in procession and marched down the middle aisle, out of the great doorway, under the English and American flags, to the churchyard, and so around the north wall to the vestry room. At the north door the clergy halted and opened, when the prelates passed through, and the procession closed after them. Then the chimes again rang out joyously, and the people separated, some to the Battery, some to other points, to witness the procession.

### THE RECEPTION OF THE GUESTS.

The first part of the programme drawn up by the Common Council Committee for the great celebration was the reception of Mr. Cyrus W. Field, the officers of the Niagara, the Gorgon, and the Indus, at the Battery. According to the official announcement this was to take place at half-past one o'clock, but long before that hour a crowd numbering at least fifteen thousand persons, of all ages and conditions in life, were assembled around Castle Garden and in the streets adjacent to the Battery. This immense congregation, finding that it was yet early in the day, concluded there was nothing for them to do but to stick by their places and gaze at the shipping. The United States frigate Sabine lay at anchor within three hundred yards of the Battery; a large American ensign floated at her peak, a small one at the mizzen, still another at the main, and the English jack flaunted at the fore. Her guns were run out ready for a salute. Those who did not know better regarded her with great admiration under the impression that she was the frigate Niagara, as the Committee of Arrangements had announced that this heroine of the sea and of the occasion would be there. With the arrival of the military the thousands already on the ground were squeezing into every nook and corner. The sun poured down its intensest rays, the soldiers sweltered, men wiped the perspiration from their brows, and loafers elbowed nurses with babes in arms, to get possession of the few shady patches that could be found. About the only comfort there was consisted in feasting the eyes upon the sight of the supposed Niagara, and the lines of the Sabine and the genius of Steers were loudly praised, until it got about that they were sold and that the Niagara still lay in tar and *dishabille* at the Navy Yard.

Still, though that was not the Niagara, everybody wondered when the Niagara would appear according to the unerring announcement of the programme. Those posted, however, were no less anxiously looking for the English officers, Captain Hudson and Mr. Field, and no less in a quandary which way to look for them. At half-past one P. M. a salute from the Sabine announced that somebody was coming. The steam tow-boat Isaac P. Smith was observed coming down the North River towards the Battery; the extra quantity of bunting displayed indicated that somebody unusual was on board, and, when she fired a salute from her forward deck, it was made sure. She stopped alongside an emigrant barge, and the heroes of the Atlantic Cable appeared at the gangway.

As Mr. Field, walking arm in arm with Captain Dayman, and followed by Mr. William E. Everett and the officers of the Gorgon and Indus, stepped ashore, the Sabine fired a thundering salute of twenty-one guns, and the crowd cheered long and loudly.

In consequence of the fatigues attendant on the discharge of the duties of Chairman of the Joint Committee, and the necessity of his presence elsewhere, Alderman McSpedon deputed Warren Leland, Esq., to tender to these gentlemen an appropriate welcome on behalf of the Committee of Management and the municipal authorities generally. Shortly before the arrival of the guests Mayor Tiemann, Peter Cooper, Wilson G. Hunt, Professor Bache and several other gentlemen, presented themselves at Castle Garden and subsequently took part in the reception.

Mr. Leland performed the duty assigned to him with his customary urbanity and good taste. These ceremonies over, the guests were conveyed to carriages, which had been in readiness for their use from an early hour, and were conveyed to the Metropolitan Hotel, where they were received by Alderman McSpedon in a few words of cordial welcome. He expressed his pleasure in having the opportunity to extend to them the hospitalities of the city, as a slight token of appreciation of their great and distinguished services in the advancement of a scientific enterprise, tending to the diffusion of the principles of a common brotherhood.

The following are the names of the British officers:

GORGON.

Captain Dayman, R. N.

Hart Gimlette, M. D.

Lieutenants Count Viscomte, B. Mitchel, T. B. Butler Paymaster A. T. N. Roberts.

INDUS.

Captain William Ross Hall, R. N.

Lieutenants Hugh Davis, L. A. Bell, Joshua C. Cole, A. H. Trainor, P. C. Johnstone and A. T. Kings.

Captain Hudson was not among those who were on the steamer, but he was at the services in Trinity Church.

Of Captain Dayman, the commander of the Gorgon, it may be proper to state, that to him was intrusted the surveying of the telegraph plateau, and by him were Lieut. Berryman's soundings corroborated. It was Captain Dayman who guided the Niagara from mid-ocean to Trinity Bay, and he is entitled to the highest praise for his admirable navigation. There was one gentleman whose absence was much regretted—Mr. James L. North, First Lieutenant of the Niagara, who kept watch and watch with Captain Hudson during the laying of the Cable, but whose name has been entirely forgotten in the praise which has been so lavishly bestowed elsewhere. The only officers present at the celebration were Mr. J. C. Eldridge, Purser of the Niagara, Lieutenant Gherardi, and Mr. Farren, the Chief Engineer, while Lieutenants Todd, Guest, Webb, Macauley, Chief Engineer Follansbee, and Drs. Green, Grinnell, Hay, and Lieutenant Boyd, of the marine corps, were absent.

The officers of the Gorgon, above-mentioned, formed a part of the procession that walked with the Cable up the hill that leads from the landing place in the bay of Bull's Arm to the telegraph station, at Trinity, N. F.

The Indus is the flagship of Admiral Seymour, and, at the time of which we write, was stationed at Halifax.

Having thus disposed of the proceedings accompanying the reception of these gentlemen, we will return to the City Hall, where, in the Mayor's Office, the guests of the Corporation were to assemble, and from which they were to enter the procession as it passed through the Park on its way to the Crystal Palace.

Policemen in uniform stood outside the door leading to the

Mayor's apartments, and prevented an early rush of visitors inside. Victims of ticket swindlers, faro banks, and mock auctions—wives, whose eyes had been blackened and heads damaged by their brutal husbands—and representatives of the usual daily heterogeneous class of complainants—presented themselves to have their wrongs righted. They were all sent away, however, and told to come again. The injunction to depart was obeyed cheerfully upon their being told that it was Cable Day. At 12 M., the Mayor's main office was well filled. The visitors were illustrious characters, and animated were their mutual greetings, the introductions and conversations. The Rev. Dr. FIELD, father of the chief hero of the day, occupied an arm-chair by the side of Lord NAPIER. Although nearly seven decades of years had passed over the head of the venerated and honored clergyman, the glow of health was still perceptible upon his cheek, and a genial, wholesome look radiated from his countenance and undimmed eyes. He entered zealously into conversation with Lord NAPIER; and the representative of Her Majesty's Government responded with his wonted ease and freedom. Bishop HUGHES sat *vis-a-vis* with the Rev. Dr. NOTT, the venerable President of Union College, with whom he conversed familiarly and with apparent pleasure. PETER COOPER, Esq., WILSON G. HUNT, Esq., and the Rev. Drs. ADAMS and SPRING, formed a party by themselves. Prof. BACHE, Superintendent of the United States Coast Survey, and Prof. HEDRICK, formerly of the North Carolina University, were busily chatting in one corner, and but a few feet from them stood, similarly engaged, Professors TORREY and WEBSTER. BENJAMIN H. FIELD, a brother of CYRUS, and JULIUS W. TIEMANN, a brother of Mayor TIEMANN, formed a talking duet. Mayor TIEMANN occupied his time in con versation with Mr. ARCHIBALD, British Consul at this port.

At a few minutes past one o'clock, the Mayor and several other gentlemen withdrew, for the purpose of taking part in the reception of the heroes of the Cable, at the Battery —a portion of the proceedings which has already been described. They returned, with their guests, in time to join in the procession, the principal features of which we shall now proceed to narrate.

THE PROCESSION.

The procession was organized according to the programme, and began to move at the pre-arranged hour. The display was one of the most imposing which the city has ever witnessed. From every street, crossing and running into Broadway, great crowds continually poured into the chief artery of the city, like so many rivulets adding their tributary streams to the mighty torrent, which gathered in bulk and force every moment until the capacity of the available area was taxed to the utmost. The march up Broadway to the Park was comparatively unimpeded, but the density of the crowd interfered at times with the movements of the military as they deployed into line, and caused some delay. When the head of the procession had reached Park row, to cross in front of the City Hall, further advance was found to be impracticable. All sorts of vehicles were jammed together at the corners of the streets, every one of them filled with motley groups on the tiptoe of expectation for the advancing *cortege.*

Long before the hour of two, at which time the procession ought to have appeared in the Park, every available spot for a sight was densely thronged with men, women and children.

After passing the Park, which occupied several hours, the procession continued its march up Broadway. The sidewalks presented a dense mass of human beings whose cheers were deafening. The windows, the house-tops, the balconies, on either hand, were filled with ladies, many of whom, in the enthusiasm of the moment, cheered also, and waved their handkerchiefs as the military passed. Regiment after regiment received its ovation from the excited crowd, and the company from Canada was saluted with an extra salvo of cheering.

Away up Broadway, as the procession moved on, the gratulations of the populace increased, and from balcony to balcony and from window to window the signal was taken up, while ten thousand handkerchiefs fluttered on the breeze.

The enthusiasm of the people never flagged for one moment; but after the chief State regiments had passed on it abated somewhat. The arrival of the carriage containing Cyrus W. Field, Esq., caused the smouldering flame to burst out afresh and with tremendous vigor. To the enthusiastic reception accorded him this distinguished gentleman modestly bowed, standing in the carriage with his hat in his hand all the while. The British naval officers —Captain Dayman, of the Gorgon, and the other officers of Her Britannic Majesty's Navy—were also similarly received and acknowledged the compliment with graceful inclinations of their heads. Captain Hudson, of the United States steamer Niagara, was loudly cheered.

The other portion of the procession that came in for a very particular manifestation of enthusiasm was the large coil of the Atlantic Telegraph Cable, which was neatly wound round a pyramid on a car drawn by six horses, gaily caparisoned with flags and plumes. This important part of the display was confided to the care of the brave sailors of the Niagara, and the handsome team by which it was drawn was furnished gratuitously by the American Express Company.

The printing press, which was kept working in the procession, the sewing machines, and the great variety of other articles exhibited, received the applause they so well deserved.

We now come to speak of the procession in detail.

To the military was given, as a matter of course, the right of the line. In accordance with the programme, they assembled at the Battery at one P.M., and, after the reception of Mr. Field, Mr. Everett, Captain Dayman, and the other British officers, filed up Broadway. Previous to this, however, the Montreal company was reviewed. This body was under the command of Captains Stephenson and Ogilvie. They were attired in blue frock coats, blue pants, and heavy fur caps. The ensign carried the Union Jack and the Stars and Stripes alternately. As the Seventh Regiment marched passed the Canadians, who saluted them in military style, the spectators raised a loud and hearty cheer.

The troops, having all marched within the gates, were put through a series of evolutions, and, everything being now ready, the First Division, at half-past two o'clock, commenced the march up Broadway, in the following order:

First, there was a detachment of sixty policemen, walking two abreast. Then came a troop of cavalry, as an escort to the Grand Marshal, Brigadier-General Yates. The Special Aids followed, consisting of these gentlemen, General W. L. Morris, Captain John Calhoun, U. S. N.; Colonel H. P. Martin, Major George E. Baldwin, John B. Rich, General Paul N. Spofford, Colonel T. F. Peers, Colonel Edward Satterlee, Major S. M. Alford, Ambrose K. Striker.

### FIRST BRIGADE.

Commanded by Brigadier-General Charles B. Spicer.
Brigade Staff.
Seventy-first Regiment—Col. Abram H. Vosburg.
Second Regiment—Col. Henry Robinson.
First Regiment—Lieut.-Col. Spencer H. Smith.
Third Regiment—Col. T. Brooke Postley.

### SECOND BRIGADE

Commanded by Col. Edward Hincken.
Brigade Staff.
Fifth Regiment—Col. Christian Schwarzwaelder.
Sixth Regiment—Lieut.-Col Samuel K. Zook.
Fourth Regiment—Lieut.-Col. Daniel W. Teller.

### THIRD BRIGADE.

Commanded by Brigadier-Gen. William Hall.
Brigade Staff
Seventh Regiment—Col. Abram Duryee.
Eighth Regiment—Col. George Lyons.
Fifty-fifth Regiment—Col. Eugene Legal.

### FOURTH BRIGADE.

Commanded by Brigadier-Gen. John Ewen.
Brigade Staff.
Tenth Regiment—Col. William Halsey.
Twelfth Regiment—Lieut.-Col. Henry A. Weeks.
Sixty-ninth Regiment—Col. James R. Ryan.
Eleventh Regiment--Col. Homer Bostwick.

There were, in all, about 6,000 men in the ranks.

The second division met in Greenwich street, the right resting on Battery place. Col. Joseph C. Pinckney was Marshal, aided by Col. Moses E. Crasto and Rufus E. Crane. It appeared in the following order:

First came the Sergeant-at-Arms of the Common Council followed by the Committee of Arrangements, in carriages

Then Mayor Tiemann and Cyrus W. Field, in a carriage drawn by four horses.

Next came Captain Hudson, of the Niagara, and Mr. Archibald, the British Vice-Consul at the port of New York.

Captain Dayman, of the Gorgon, and Mr. W. E. Everett, with officers of the City Government, came next.

The next carriage contained Lord Napier, the British Minister, the Most Rev. Archbishop Hughes and the Rev. Dr. Nott. These eminent personages attracted much attention and were frequently hailed with cheers.

The orator of the day, David Dudley Field, the officiating clergy, and Peter Cooper, came in the next carriage and were repeatedly cheered by the crowd.

A succession of carriages then followed, containing Aldermen and other city officials. With them were the officers of the Niagara and Gorgon, and the officers of the New York, Newfoundland, and Atlantic Telegraph Company.

Mayor Powell and representatives of the Common Council of Brooklyn.

The Commissioners of Emigration.

The Board of Ten Governors, in full.

Sheriff Willett and deputies, County Clerk R. B. Connolly, Registrar W. E. Miner, Coroners Connery and Gamble, and representatives from the Board of Supervisors, Central Park Commissioners and Police and Health Commissioners.

The Mayor and Common Councilmen of Hoboken, Jersey City, Hudson City and Harsimus, were present in seven large omnibuses, which were decorated with banners, flags and devices of various kinds.

On the top of the principal stage was a representation of the Niagara paying out the Cable, on either side of which were the following inscriptions:

"PERSEVERANCE MUST PROSPER."

"UNION OF STATES AND COMBINING OF NATIONS."

Among the other inscriptions decorating the stages were the following:

"THE ILLUSION DISPELLED THAT NEW JERSEY IS BEHIND THE AGE."

"England and America—Once Separated by War, Now United by Lightning."

"A mighty work has now been done—
All praise to Massachusetts' son;
His name resounds from shore to shore,
And echoes through the ocean's roar."

"Let the Eagle scream."

"Hurrah for Bergen enterprise,
Hurrah for Yankee skill—
One navigates the ocean's depths,
The other—Bergen hill."

"Nations are now exchanging Sentiments by Lightning Speed."

"New York City has her Cyrus,
And Jersey has her Jake,
With the tallest line of stages
In little Jersey State."

"The Ladies—The Mothers of Inventors."

"When American Genius spans the Ocean, what will it not do?"

"Franklin and Morse—One drew the Lightning from the Clouds, the Other drew it through the Ocean Depths."

"May the Atlantic Cable hold the two Continents together, and make Peace perpetual."

"The Lightning tamed by Morse, the Ocean chained by Field."

"Preserve the Union—Trinity and Valencia."

"FRANKLIN'S MEMORY FRESH AND SWEET."

"MARRIED—ENGLAND AND AMERICA, BY THE CABLE'S CHAIN."

"CAPTAIN HUDSON, THE PRIDE OF THE NATION; THE NIAGARA, THE GEM OF THE OCEAN."

These inscriptions on the Jersey stages were the theme of much favorable comment, and attracted great attention. Indeed, the display of the Jersey contingent was, in every respect, most creditable. The grand feature, however, of this division was a large car of the American Express Company, drawn by ten handsome bay horses, on which was mounted a telegraph instrument and a large section of the Atlantic Cable, coiled up in the form of a pyramid, surmounted by a liberty cap and the American flag. On either side of the pyramid stood a jack-tar, one of whom grasped the American and the other the British colors. An operator was working the Cable with one of Hughes' instruments, and throwing the dispatches right and left. The following is a copy of the dispatch:

"SEPTEMBER 1ST. AMERICA TO EUROPE, GREETING: PEACE ON EARTH AND GOOD-WILL TOWARD MEN."

Behind this car came the crew of the United States ship Niagara, bearing before them a model of their splendid ship. These hearty fellows were vehemently applauded.

Another large car, belonging to the Adams' Express Company and drawn by twelve fine chestnut-colored horses, richly caparisoned, followed the crew of the Niagara. It contained one of Hughes' telegraph instruments and a banner, on which was inscribed:

"AMERICAN TELEGRAPH COMPANY."

Then came an immense car, drawn by eight horses, on which were mounted three printing presses, namely, one of Messrs. R. Hoe & Co.'s single cylinder presses, a card press, and a queer old relic, in the shape of a wooden press, more than a hundred years old, with the primitive buckskin balls for inking apparatus. Printed sheets were thrown off by the respective presses on this car, as the procession moved along.

Among the printed matter worked off by the press, and scattered among the crowd, were the ode written by Mrs. Stephens, a paper entitled "The Atlantic Telegraph," another purporting to give "A Brief History of Printing," and the Queen's and President's messages, all of which were very creditable specimens of printing. The contrast between the press as it was one hundred years since and the printing press as it is now, was brought before the public eye vividly. Four pretty little girls, dressed in white, with red trimmings, helped to distribute the printed matter.

The New York Typographical Society, a time-honored Association, followed immediately after the presses, bearing a banner on which was inscribed:

"NEW YORK TYPOGRAPHICAL SOCIETY, INSTITUTED, JULY 4, 1809."

On the reverse side was the legend:

"THE LIBERTY OF THE PRESS IS ESSENTIAL TO THE PRESERVATION OF FREEDOM."

There were about two hundred members of the Association out.

The Third Division met in Beaver street. It was officered by Lieutenant-Colonel Marshall Lefferts, as aid to the Grand Marshal, assisted by Major J. R. Pinckney and R. H. Shannon.

In this appeared:

*Consuls.*—D'Aguier, of Brazil; Don Santos, of Portugal; in a carriage from which waved the Brazilian flag; L. H. Hatfield, of port of Bombay, India.

*Militia Officers off Duty.*—Lieutenant F. W. Obernier, of the 72d Regiment; Lieutenant Laidlaw, of the 14th; ex-Sergeant-Major Van Tassel, Sixth Regiment; Captain Taylor, Fourteenth.

*Officers of Telegraph Companies*—In United States Express wagons, drawn by six horses, containing the managers and employés of the Morse and House Telegraph Societies.

The Mercantile Library Association was represented in this division by about two hundred intelligent-looking young men.

On a rich silken banner which they bore was inscribed:

"MERCANTILE LIBRARY ASSOCIATION OF THE CITY OF NEW YORK."

"FOUNDED 1820."

The Fourth Division was under the direction of General Lucius Pitkin, aided by Captain W. W. Leland and J. W. Downing, Esq. It was composed of the Second Company, Washington Continentals, Captain Lansing, thirty-five muskets, acting as escort to the Veterans of 1812, who appeared on the ground, numbering one hundred men, under the command of Colonel H. Raymond. Colonel Fox, of the Veterans, who claims to have killed, in the war of 1812, the biggest Indian ever seen this side of the Rocky Mountains, paraded a sword that was worn by Commodore Perry in the naval battle fought on Lake Champlain. This sword was given to a nephew of Perry's, who died while on a station on the Mediterranean. The sword came into the possession of Thomas M. Carr, the Consul to Morocco, who subsequently presented it to its present owner.

St. George's Society, in three carriages; one containing the President, the two Vice-Presidents and Secretary. This carriage was enveloped in the large banner of the society. The other carriages contained members—each vehicle draped with the English flag.

St. Andrew's Society, in four carriages, the first containing the Vice-President, Managers, and Secretary. The officers all wore badges. On the top of their carriage was seated the standard-bearer of the Society, carrying the flag of St. Andrew. The other carriages contained members of the Society.

Friendly Sons of St. Patrick, in one carriage, containing the officers of the Society. On top of the vehicle floated the green flag of Erin.

Italian Benevolent Society, marching five abreast. This Society carried a magnificent banner eighteen feet in length and eight feet in breadth, on top of which was a representation of Columbus discovering the New World, with the inscription underneath:

"COLUMBUS UNVEILING THE NEW TO THE OLD WORLD."

Beneath this was a representation of the laying of the Cable, with the inscription below of

"CYRUS W. FIELD JOINING THEM BY THE ATLANTIC CABLE."

On the left side, on top, were the names of Franklin, Morse, Hughes; on the right side, Galvani, Volta, Bonnelli. The painting was surrounded with thirty-one silver stars. In addition, the Society carried a large Italian flag.

The Fifth Division, which formed in Cedar street, was commanded by Colonel John W. Stiles, aided by H. E. Phelps and A. H. Van Pelt, Esqs.

Foremost and most prominent in the line was the National Cambrian Standing Committee, a Welsh organization, which, with thirteen harps, representing their thirteen native counties, were located on an immense truck which was decorated with ribbons. Thirteen skillful harpers performed national airs on the harps during the march. On each side of the wagon, the platform of which was twenty-six feet long, a banner was displayed. The stand banner contained the following inscription in Welsh, over which a terrible-looking red dragon was spreading its wings

"TNJOPVK—JNJOPVK—NYNT."

which signifies:

"MY LANGUAGE—MY COUNTRY—MY NATION."

This truck contained also a piece of Telegraph Cable suspended on a wheel, as if it were in the act of being paid out, while the truck was moved by ten horses.

The New York Caledonia Club, a Scottish National Association, followed. The members, one hundred in number, turned out in their national costume, bare legs and kilts, with eagles' plumes in their caps, and were followed by half a dozen bag-pipers. The standard they bore was a crown of flowers, on its top was a liberty cap, and below was the inscription:

"WE'RE BRETHREN A'."

The next most interesting feature of the Fifth Division was, undoubtedly, the model of the engines of the United States frigate Niagara, which took so prominent a part in the work of submerging the Cable. Owing to the general desire to become acquainted with everything connected with this noble vessel, large crowds were congregated around the model closely inspecting it.

The Cadets of Temperance, a Society of young men, turned out in eight sections and mustered about four hundred. The sections were the Mount Vernon, Mechanics, Williamsburg, Enterprise, Excelsior, Mercantile, Lawrence and Ashland. They were accompanied by a band of music, and carried several banners with appropriate inscriptions.

The German Workingmen's Clubs, composed of deputations from the Tenth, Eleventh, Seventeenth, Eighteenth and Twentieth Wards, turned out about one hundred strong, under the command of Mr. Danke, and were remarkable for their orderly appearance. They were accompanied by several bands of music, and carried several banners and American flags. On one of the banners was a representation of two hands clasping in friendship and the words:

"ARBITER VEREIN."

The Randall's Island boys, under the direction of Mr. Tappen, brought up the rear of the Fifth Division. They were dressed plainly but neatly, in light blue coats, white pants and check caps, mustering one hundred and twenty strong, under the command of their young captain, Edward Ryan. The boys marched with all the regularity and precision of old soldiers.

They walked eight abreast, to the music of fife and drum, and carried flags and banners, one of which was presented to them by a philanthropic Society in Philadelphia, and is said to have cost one hundred dollars. On one side it bore a spirited representation of "Young America," underneath which were the words "Boys of '76;" the other side had the following inscription:

"Presented to the male orphan children of Randall's Island, by Joseph G. Steel, George Moore, Charles H. Geehr, Harrison G. Clark, Joseph Delavan, A. W. Paynter, H. F. Schellenger, P. H. Beck, William W. Long, Philadelphia, Pa."

The Sixth Division was organized under the direction of Colonel Thomas F. Devoe, one of the Grand Marshal's aids, assisted by Captain Samuel F. Patterson and Mr. D. J. Levy.

The order of procession was as follows:

American Institute.

Mechanics' Institute.

Officers of the United States Government.

American Telegraph Co., with Hughes' printing instrument.

Inventors of sewing machines.

Sewing machines on a car, and operatives at work.

Pianoforte makers.

Pianofortes on truck, with performers.

Sewing machines on a car, with operatives at work.

Model of United States steam frigate Niagara.

Daguerreotype and photographic artists.

Envelope manufacturers.

The most imposing feature of this division was the display of sewing machines. One establishment was represented by a large and beautiful canopy, fifteen feet long, eight feet wide and eighteen feet in height, erected on a large four-wheeled wagon, drawn by six richly-caparisoned white horses decked with flags and plumes. The canopy was intended to represent a drawing-room scene, it is to be supposed, from the fact that the flooring was elegantly carpeted, and several luxurious sofas and chairs were arranged in various positions in the most tasteful order. In the centre of the car was stationed a fine sewing machine in operation, attended by a beautiful young lady. On the top of the canopy, which finished in a beautiful dome, was an American eagle bearing a floral wreath in his beak. The canopy was hung all around with costly lace curtains and bore several inscriptions.

Another sewing machine factory also made an excellent demonstration. This was simply composed of an extensive platform, erected on a large wagon, drawn by twelve horses, six gray and six black. The platform was divided by a transparent screen into two partitions ; one containing three young ladies sewing with the needle, and intended to represent the old system in vogue before the invention of the sewing machines. Above this partition was the inscription :

"THE SONG OF THE SHIRT."

The girls in this partition, it is stated, were given strict orders to look as interestingly miserable and careworn from "the stitch, the stitch, the stitch," as they could possibly assume ; but spite of themselves and their instructions, they wore terribly bewitching smiles, and appeared quite as merry as any of their sweet sex who turned out in such formidable array as spectators. In the back partition four girls were seated, tending a sewing machine. Above this partition were the words—

"TRIUMPH OF SCIENCE."

One of the greatest objects of attraction in the whole procession was probably a small wagon, drawn by two beautiful little Mexican ponies, and driven by a little man not over two feet in stature.

An interesting display was made by an ink-making establishment. It consisted of a platform, placed on a wagon drawn by four horses, with a large black ink bottle twelve feet high, numerous boxes of ink, and twenty-five men from the manufactory. The mottoes on the car were :

"NOT FOR A DAY, BUT FOR ALL TIME."

And

"FESTINA LENTE."

The photographic department was represented only by a wagon, drawn by two horses. On the wagon was a large double patented camera, raised about twelve feet from the ground, and surmounted by a bust of Benjamin Franklin, held in its position by a boy seated beside it. At each side of the camera, and just beneath it, was a long white scroll, on which were the inscriptions:

"GOD'S POWER! BEHOLD THE SUBJUGATION OF HIS ELEMENTS!"

"ENGLAND." [Two hands clasped.] "AMERICA!"
"The inventions of man—science and art—
Elevate labor and refine the heart."

The pianoforte makers turned out handsomely. One concern furnished a large platform on wheels, which was surrounded by an iron railing and decorated with four large flags and twenty-eight smaller ones. On this platform was placed a beautiful pianoforte, at which a performer was seated and played at intervals. Another factory had one of their pianos on a car, drawn by four horses. The instrument was surrounded by the following inscriptions:

"HIS LIGHTNINGS ENLIGHTENED THE WORLD."

"THY WAY IS IN THE SEA,
THY PATH IN THE GREAT WATERS."

The envelope makers turned out with a large triumphal car, drawn by three fine horses and carrying an envelope machine in operation, tended by two girls. Around the top of the framework of the platform was, at either side, a representation of the Cable extending between Valentia and Trinity Bay, one of the ends entering an envelope.

Following this was another car, drawn by four horses, carrying some agricultural implements.

Next came Joseph Millward's braiding machine, in operation and tended by several young girls. This was carried on a wagon, drawn by two horses.

A large builder's wagon, containing workmen and their tools, and a car, containing a pretty extensive display of tin and metal roofing, brought up the rear of the Sixth Division.

The Seventh Division formed on Dey street, the right resting on Broadway. It was under the command of Alfred A. Phillips, one of the Grand Marshal's Aids, assisted by John J. Ogden.

The first organization in this division was the New York Turnverein. It was preceded by the Turner Rifles, numbering fifty men, under the command of Captain Billou. The Turnverein numbered five hundred men, all dressed in the white uniform, with red cravats and black hats. It was under the command of Captain Sanger. The cortege was preceded by a band, consisting of sixty musicians.

Next came the Allgemeine Sangerbund, under the command of Captain Vogol. This party also numbered five hundred men, and were dressed in a black uniform. Both the Turners and Sangerbund Society carried an immense number of large flags of all colors. The effect was quite imposing.

Next came the Independent Singing Society, some two hundred and fifty in number, and carrying flags similar to those in the Societies above mentioned.

The Columbia Song Society, a German organization, followed, and presented a creditable appearance.

The Eighth Division was under the direction of General Henry Storms, Aid to the Grand Marshal, assisted by John Benson, Esq., and shortly after one o'clock formed in John street, the right on Broadway. It was led by

The Butchers of the cities of New York, Brooklyn and Jersey City, numbering about four hundred. They were on horseback, and wore linen aprons and over-sleeves. Each member had pinned to the lappel of the coat a neatly printed portrait of Mr. Field, with the words :

"BUTCHERS."

"PORTRAIT OF CYRUS W. FIELD."

Many had also attached to it portions of the Cable made into chains. Following the members came a large wagon, drawn by six horses and beautifully decorated with flags. A huge stuffed ox was placed on it, and underneath the ox was a diminutive steer. Around the wagon were the following inscriptions :

"OX WASHINGTON.
LIVE WEIGHT 3,400 LBS."

"THE ATLANTIC CABLE—A RAY FROM THE SUN OF LIBERTY, DESTINED TO ILLUMINATE THE GLOOMIEST RECESSES OF THE WORLD."

"THE CABLES OF TYRANNY ARE BLIGHTED BY THE CABLES OF INTELLIGENCE."

Numbers of cars, wagons, &c., followed, carrying the members of the Societies of Jersey and Brooklyn. They were all gaily decorated, and the appearance of the entire body was highly creditable.

"Agricultural interest, specimens of cattle, horses," &c., followed. A large wagon, drawn by six horses, and in which were several sheep adorned with ribbons, was the first. Around it were the following inscriptions:

"THE FIELDS OF AGRICULTURE
PAY HOMAGE TO THE FIELDS OF SCIENCE."

"THROUGH THE CLEFT WATERS OF THE SEA
RESOUNDS EARTH'S ANTHEM—MAN BE FREE."

A wagon of considerable size, drawn by six horses, and containing sheep, oxen, dogs and all the animals of the farm-yard, was the next. Suspended from the outside were all descriptions of vegetables, corn, cabbage, beets, carrots, &c., of unusual quality.

This display of agricultural products was loudly cheered as it proceeded, and elicited high encomiums, "The Van Brant packing horse" had a beautiful wagon, decorated with all kinds of flowers and ribbons.

The Croton Mills exhibition consisted of a large truck, drawn by six horses and piled up with bags of flour. This was followed by many carts and trucks, loaded with large millstones and sacks of the "Milton Brand." All the cars were richly decorated. Following came a horse, with a bag of wheat thrown across and ridden by a youth in rustic attire, with an inscription:

"GOING TO THE MILL IN THE OLDEN TIME."

The Croton Mills and Metropolitan Mills had also several trucks with specimens of flour and sacks of wheat.

A number of Shetland ponies closed this portion of the procession.

The Workingmen's Association, numbering some three hundred members, followed, and a delegation from the Palace Gardens, in carriages. After them came the Longshoremen's Society, which closed the division. They mustered strong and looked admirably. As they walked up Broadway they were loudly cheered by the spectators.

The Ninth Division was commanded by Captain G. S. Mott, assisted by Lieutenant Macintosh and D. W. C. Rice, Esq.

First came the cartmen of the city of New York. They numbered upwards of two hundred; their carts were all decorated with flowers and neatly painted for the occasion. They were respectably attired and presented a fine appearance. Many of the members had their wives and children with them. The children were dressed in white and covered with ribbons and garlands. The next in order were—

The milkmen of the city of New York. There were nearly two hundred milk carts in the procession, bearing the well-known "Pure Orange county milk." Some of the carts were newly painted and the milk cans were all highly polished. The horses were covered with flowers, and on some of the cars inscriptions were suspended. One:

"VICTORIA, BUCHANAN AND FIELD,
MAY THEIR SHADOWS NEVER BE LESS."

On another:

"WITH LOUD HUZZAS NOW LET THE WELKIN RING,
COLUMBIA'S GOT BRITANNIA ON A STRING."

Moulding and planing machine of the New York Mill, Thirty-seventh street, in full operation, followed, and was loudly cheered. It was drawn on a capacious truck by six horses, all of whom were decorated with flowers and ribbons.

Brewers with their horses and wagons followed, and were a great feature of the procession. The New York steam brewery had several trucks, and on one a complete steam apparatus at work. Displayed in a most prominent position was the following:

"NO COCCULIS OR CANNABIS CONTROLS OUR POWERS—
BUT BOUNDLESS FIELDS OF BARLEY MALT ARE OURS."

On the reverse side,

"THE TWO EPOCHS OF THE AGE—ELECTRICITY AND STEAM ALE."

Nearly twenty wagons, drawn by eight and ten horses, all decorated with flags, &c., appeared in this Division.

The electric oil charger and a camphene wagon gayly dressed with flags, &c., succeeded, and were followed by

Coal dealers with specimens of American coal.

Among these was a beautiful cart drawn by fifteen horses, with fine specimen of cannel coal. The horses were all decked out with flowers and flags.

Gas generating and cooking ranges, drawn on trucks, came next.

Then followed a perfect cooking apparatus, drawn on a large truck, with the following inscription:

"THE BIRTH-YEAR OF THE TWO GREAT WONDERS OF THE AGE—THE ATLANTIC TELEGRAPH,* AND THE NON-EXPLOSIVE GAS GENERATING COOKING RANGES."

Camphene and alchohol distillers, and the new steam carriage of Mr. Dudgen, joined this division of the procession.

The printing ink manufacturers followed, and made a creditable display.

The coopers, saddlers, sewing machine makers, with the implements of each trade at work, came next, and after them a number of safes, of all sizes, drawn by eight powerful horses, and beautifully decorated.

Platform scale manufacturers and gutta percha life-boat manufacturers followed.

A gutta percha life-boat was drawn on a lofty car by six horses. On the side of the boat, which was decorated with flags, was the inscription:

"ON OCEAN'S BOSOM SAFELY FLOATS
LARCHER'S GUTTA PERCHA BOATS."

A patent bread and biscuit machine, drawn by six horses, succeeded.

This division, being the last, was closed up with a large number of citizens, walking six abreast, who accompanied the procession to the Crystal Palace, which it reached about six o'clock.

## RECAPITULATION.

The procession took one hour and forty minutes to pass, omitting stoppages.

Annexed is a recapitulation of the number in each Society that turned out:

| | |
|---|---|
| The Military | 6,000 |
| Mercantile Library Association | 300 |
| Italians | 200 |
| Foreign Residents and Societies | 300 |
| Telegraph Operators | 100 |
| Typographical Society | 200 |
| Printers | 200 |
| Sailors | 300 |
| Expressmen | 100 |
| Officials and Guests | 100 |
| Veterans of 1812 | 100 |
| Caledonia Club | 150 |
| Cambrian Society | 100 |
| Cadets of Temperance | 400 |
| Working Men's Clubs | 100 |
| Randall's Island Boys | 120 |
| On wagons, (estimate) | 400 |
| Staten Islanders | 100 |
| Federal Officers | 25 |

| | |
|---|---:|
| Butchers of New York, Brooklyn, and Jersey City | 400 |
| Workingmen's Association | 300 |
| Longshoremen's Society | 300 |
| Cartmen of the city of New York | 200 |
| Milkmen of New York | 300 |
| New York Turnverein | 460 |
| Algemeine Sangerbund | 370 |
| Independent Singing Society | 235 |
| Columbia Song Society | 200 |
| Krakehlia Dramatic Society | 115 |
| Band | 25 |
| Cavalcade of Equestrians' Institute of Williamsburg | 210 |
| Smaller societies, not enumerated | 3,000 |
| Total in line about | 15,410 |

PROCEEDINGS AT THE CRYSTAL PALACE.

It had been arranged that the procession should arrive at the Palace at four o'clock, or as near to that hour as might be practicable. In anticipation of this, the doors were thrown open at two, and immediately the building began to fill. Those to whom green tickets had been given entered by the main door, on Sixth avenue; those who obtained white tickets were admitted through the side entrance, on Fortieth street, and had access to the galleries in the vicinity of the platform, which was erected in the nave, and extended backwards from the centre to the eastern window. The platform proper occupied but half of this space. Midway between the fountain (which is directly beneath the dome, and but a few feet from the front of the platform,) and the eastern end of the nave, commenced the first of a tier of seats which, rising by regular gradations from the stand, terminated at the farther wall and on a level with the galleries. These were set apart for the band, the members of the New York Harmonic Society, the Harmonic Society of Brooklyn and Williamsburgh, the Mendelssohn Union and the Liederkranz Society. The conductor's baton was intrusted to Mr. George F. Bristow, under whose direction the musicians acquitted themselves most admirably. The entire platform was handsomely carpeted, and both the platform and the gallery, improvised for the orchestra, were draped with the "Red, White and Blue." Above were suspended the English and American colors, and from the front of the north-western gallery depended a large green banner, bordered with golden shamrocks and displaying in the centre an Irish harp—"the sunburst of Erin." Beyond these, the Palace was destitute of any attempt at decoration.

At five o'clock the musicians entered and took their seats on the raised benches. Soon after a number of the invited guests appeared, among whom were Lieutenant M. F. Maury, of the Observatory at Washington, Rev. Dr. Adams, of the Madison Square Presbyterian Church, Rev. Dr. Ferris, Chancellor of the New York University, and Rev. Alfred J. Dayman, (brother of Captain Dayman, of the Gorgon), of the Fiftieth street Catholic Church.

At a quarter before six o'clock the roll of drums and a loud cheer outside proclaimed the arrival of the procession, and Mayor Tiemann, escorting Cyrus W. Field and followed by Captain Hudson, Captain Dayman, Mr. W. E. Everett, the officers of the Niagara, Indus and Gorgon, Most Rev. Archbishop Hughes, Lord Napier, British Minister at Washington, Mayor Powell, of Broooklyn, and other invited guests, entered by the door on Forty-second street, and were received with hearty applause, which was repeatedly renewed as they ascended the platform. Immediately after these dignitaries had taken their places the crew of the Niagara entered, bearing banners, flags and models of their ship and the Agamemnon. As soon as they had reached the front of the platform one of them called for "three cheers for Captain Hudson," and all cheered lustily. "Three cheers for Cyrus W. Field," was the next cry, and it met with a thundering response. Queen Victoria, President Buchanan, Mr. W. E. Everett, and the City of New York, were all similarly honored.

In the mean time the advance guard of the Seventy-first Regiment entered the northern gallery, and deployed into line. The evolutions of this corps were highly applauded and their presence excited no little attention. By degrees several other bodies of the military in the procession appeared and took the places assigned them. Six o'clock having at length arrived, Mayor Tiemann advanced to the front of the platform and, in a few words, announced the purpose of the demonstration. The cheers which greeted his appearance were loud and long continued.

The musical performers then rose, and rendered in magnificent style Haydn's well-known chorus:

"Achieved is the glorious work,  
Our song let be the praise of God,  
Glory to His name forever!  
His soul on high exalted reigns;  
Hallelujah! Amen!"

Rev. Dr. Adams offered prayer, after which the musicians rose, and sang the following ode:

### THE CABLE.

BY MRS. ANN S. STEPHENS.

AIR—" *Star Spangled Banner.*"

Oh, say not the old times were brighter than these,
When banners were torn from the warriors that bore them;
Oh, say not the ocean, the storm or the breeze,
Are freest or proudest when war thunders o'er them;
For the battle's red light grows pale to the sight,
When the pen wields its power, or thought feels its might:
Now mind reigns triumphant where slaughter has been!
Oh, God bless our President! God save the Queen!

Let the joy of the world in rich harmony rise.
Let the sword keep its sheath, and the cannon its thunder;
Now intellect reigns from the earth to the skies,
And science links nations, that war shall not sunder!
Where the mermaids still weep, and the pearls lie asleep,
Thought flashes in fire through the fathomless deep:
Now mind reigns triumphant where slaughter has been!
Oh, God bless our President! God save the Queen!

When the sunset of yesterday flooded the West,
Our old mother-country lay far in the distance;
But the lightning has struck!—we are close to her breast—
That beautiful land that first gave us existence!
We feel, with a start, the quick pulse of her heart—
And the mother and child are no longer apart!
For mind reigns triumphant where slaughter has been!
Oh, God bless our President! God save the Queen!

The blood that was kindred throbs proudly once more,
And the glow of our joy fills the depth of the ocean;
It thrills through the waves, and it sings on the shore,
Till the globe to its poles, feels the holy commotion!
Let us join in our might, and be earnest for light;
Where the Saxon blood burns, let it strive for the right;
For mind reigns triumphant where slaughter has been!
Oh, God bless our President! God save the Queen!

This over, Mayor Tiemann again came forward, and, amid vehement cheering, presented to Cyrus W. Field, Esq., the freedom of the city in a gold box. In making the presentation, he said:

Sir—History records but few enterprises of such "great pith and moment" as to command the attention, and, at the same time, enlist the sympathies of all mankind. In all ages warlike expeditions have been undertaken on a scale of grandeur sufficient to astonish the world; but the evils which are inseparable from their prosecution, have always sent a thrill of horror through the anxious nations. The discovery of this western continent even, the grandest event of modern times, was made by an insignificant fleet which left the shores of Portugal without attracting the notice of the civilized world. Far different has been the history of the daring and difficult enterprise of uniting the Old World and New by means of the Electric Telegraph.

From the very outset, the good, the great and the wise of all lands beneath the sun, have watched with intense anxiety, and even when doubt existed, with warm interest, every step taken towards the accomplishment of what was universally acknowledged to be the most momentous undertaking of an age made marvelous by wonderful scientific and mechanical achievements. The two greatest and freest nations of the globe by independent constitutional legislation, and by the aid of their finest ships and their ablest officers and engineers, combined together to insure success. Capital was liberally subscribed by private citizens in a spirit which put greed to the blush. The press on both sides of the Atlantic recorded the details of the progress of the undertaking with cordial interest, and secured the generous sympathies of men of all kindreds, and tongues, and nations in its behalf. You were thus fortunate, sir, in being identified with a project of such magnificent proportions and universal concern.

But the enterprise itself was no less fortunate in being projected and carried into execution by a man whom no obstacles could daunt, no disasters discourage, no doubts paralyze, no opposition dishearten. If you, to whom the conduct of this great enterprise was assigned by the will of Providence and the judgment of your fellow-man, had been found wanting in courage, in energy, in determination, and in a faith that was truly sublime, the very grandeur of the undertaking would only have rendered its failure the more conspicuous. But, sir, the incidents of the expedition and the final result—too familiar to all the world to need repetition here—have demonstrated that you possessed all the qualities essential to achieve a successful issue. It is for this reason that you now stand out from among your fellow-men a mark for their cordial admiration and grateful applause.

The city of your home delights to honor you; your fellow-citizens, conscious that the glory of your success is reflected back upon them, are proud that your lot has been cast among them. They have already testified their appreciation of your great services and heroic perseverance by illuminations, processions, serenades and addresses. And now, sir, the Municipal Government of this, the first city on the Western Continent, instruct me, who have never felt the honor of being its chief magistrate so sensiby as in the presence of this vast assemblage of its fair women and substantial citizens, to present to you a Gold

Box, with the arms of the city engraved thereon, in testimony of the fact that to you mainly, under Divine Providence, the world is indebted for the successful execution of the grandest enterprise of our day and generation ; and in behalf of the Mayor, Aldermen and Commonalty of the city of New York, I now request your acceptance of this token of their approbation. In conclusion, sir, of this the most agreeable duty of my public life, I sincerely trust that your days may be long in the land, and as prosperous and honorable as your achievement in uniting the two hemispheres by a cord of electric communication has been successful and glorious.

Mr. Field advanced to reply, amid a storm of cheers again and again renewed. He said :

Sir—This will be a memorable day in my life ; not only because it celebrates the success of an achievement with which my name is connected, but because the honor comes from the city of my home—the metropolitan city of the New World. I see here not only the civic authorities and citizens at large, but my own personal friends—men with whom I have been connected in business and friendly intercourse for the greater part of my life. Five weeks ago, this day and hour, I was standing on the deck of the Niagara, in mid ocean, with the Gorgon and Valorous in sight, awaiting for the Agamemnon. The day was cold and cheerless, the air was misty, and the wind roughened the sea ; and when I thought of all that we had passed through—of the hopes thus far disappointed—of the friends saddened by our reverses—of the few that remained to sustain us, I felt a load at my heart almost too heavy to bear, though my confidence was firm and my determination fixed. How different is the scene now before me—this vast crowd testifying their sympathy and approval; praises without stint and friends without number. This occasion, sir, gives me the opportunity to express my thanks for the enthusiastic reception which I have received, and I here make my acknowledgments before this vast concourse of my fellow-citizens. To the ladies I may, perhaps, add that they have had their appropriate place; for when the Cable was laid the first public message that passed over it came from one of their own sex. This box, sir, which I have the honor to receive from your hand, shall testify to me and to my children what my own city thinks of my acts. For your kindness, sir, expressed in such flattering, too flattering, terms, and for the kindness of my fellow-citizens, I repeat my most heartfelt thanks.

In conclusion, gentlemen, let me read you a dispatch which I have received from a little suburban village of New York; in regard to everything which takes place therein it is natural that we should feel a deep and lively interest.

Mr. Field then read the dispatch, which was greeted with the utmost enthusiasm:

"LONDON, September 1, 1858.

"CYRUS W. FIELD, *New York:*

"The Directors are on their way to Valentia to make arrangements for opening the wire to the public. They convey, through the Cable, to you and your fellow-citizens, their hearty congratulations in your joyous celebration of the great international work."

After which he resumed his seat amid tremendous cheering.

Mr. John Clancy, President of the Board of Aldermen, rose, and, in making a similar presentation, addressed Mr Archibald, the British Vice-Consul at this port, who represented Captain Preedy of the Agamemnon, and said:

Sir—The Common Council of New York, assembled with a great multitude of their fellow-citizens to celebrate the successful laying down of the Atlantic Telegraph Cable, are desirous of including in the well-deserved honors of the occasion yourself and your brother officers, the British Engineers, Electricians and the crew of the vessel under your command. While publicly expressing our admiration and gratitude to our own countrymen concerned in this wonderful achievement, we have felt it our duty that their respected associates beyond the sea, united with them in the laborious and often harassing efforts necessary to effect the great work, should share with them in our hearty recognition of their services. We regret only that we cannot, on this joyous occasion, have the happiness of tendering our grateful acknowledgments in person.

We regard it, indeed, as one of the most pleasing circumstances connected with this all-important enterprise, that it has been achieved by the united counsels, energies and means of the two great kindred nations.

As electrical science has been brought to its present highly-advanced state, by a series of profound researches, successful experiments and brilliant discoveries, alternately made on either side of the water, from the time of

Franklin down to the present day, we reflect with extreme satisfaction that this last, and all but miraculous, practical, application of the subtle element, has been accomplished by the combined scientific knowledge, mechanical skill, the united energy, perseverance and indomitable spirit of England and the United States, under the auspices of their respective Governments, and with the aid of gallant and able officers of both services. We desire especially to offer to you, in an appropriate testimonial, the high sense entertained by ourselves and the community we represent, of the cordial and efficient manner in which, with the officers and crews of the vessels under your command, you have co-operated in laying down the Atlantic Cable, and that, too, while the inherent difficulties of the undertaking were greatly augmented by stress of weather, so severe as to put the highest seamanship to the proof. We doubt not that the brave officers of the Agamemnon, in common with their brotherhood of the Niagara, have regarded the successful accomplishment of this beneficent work as a greater triumph than any to be gained by the effusion of blood. May a kind Providence, overruling all things for the honor and welfare of both countries, grant that their gallant navies may never be engaged, except as now, with generous emulation in the arts of peace.

May this auspicious enterprise, happily accomplished by the co-operation of England with the United States, have its first effect in confirming their friendly relations. While these astonishing facilities of communication accrue to the benefit of the whole civilized world, may they prove especially instrumental in establishing peace and a mutually beneficial intercourse between the two countries, on a basis never to be shaken.

For your share in bringing about the great result in which these hopes are founded, the Common Council of New York, on behalf of their fellow-citizens, now tender to you, sir, to your brother officers, to your scientific associates, and the crew under your command, our earnest and heartfelt congratulations.

Mr. Archibald replied as follows :

*Mr. Mayor and Gentlemen of the Common Council :*

Next to the consciousness of duty fulfilled, there can be no more gratifying reward than the approbation of our fellow-men. To have been considered by this, the great commercial metropolis of America, as deserving of the eloquent tribute of praise which you have awarded to the commander, officers and crew of Her Majesty's ship Agamemnon, for the part which they performed in the great enterprise of laying the Atlantic Cable, must ever be remembered by them as among their most highly prized rewards.

Intrusted, in common with their brave and gallant companions of the Niagara, with a task demanding the exercise of consummate skill, unceasing vigilance and untiring labor—all but overwhelmed by the fury of a protracted tempest in which they disdained to save their lives by the sacrifice of their precious freight, the custody of which was, indeed, their greatest peril—undaunted by repeated and most discouraging disasters—still, ever bearing in remembrance the inestimable importance of their high mission, and devoutly trusting in that overruling Providence without whose favor no work of man can be accomplished, they were at length permitted to see their arduous labors crowned with a success that has earned for them the high honors which the city of New York has now bestowed upon them.

The noble ship which it was the good fortune of Captain Preedy to command, had already won for herself, under a brave commander, high renown in fighting the battles of her country. Henceforth, her name and fame must forever be associated with a victory incomparably more glorious than conqueror or hero ever won.

Be assured, Mr. Mayor and gentlemen, that among the recollections of the all-important service in which Captain Preedy and the officers of the Agamemnon have been engaged, none are more grateful, and none will be more enduring, than the remembrance of the cordial good feeling which, from first to last, subsisted between them and their brave and skillful colleagues, Captain Hudson and the officers of the Niagara. Bearing in mind that when, on that ever memorable day, in mid-ocean, they last consigned to the bottom of the great deep that mystic cord through which is now happily established a common pulsation in the hearts of two kindred nations, they, as the representatives of their respective countries, then and there inaugurated a solemn bond and covenant of peace between them, their mutual desire and aim must henceforth ever be to rival each other only in deeds which shall conduce to the happiness and prosperity of their common countries.

Mr. Mayor and gentlemen, regretting sincerely that Captain Preedy is not here himself to participate in person in the ovation which New York has decreed in honor of those who successfully accomplished the great Atlantic Telegraph; and, unworthily representing him as I do on this occasion, I beg leave, on behalf of himself, his officers and the crew of the Agamemnon, to express to you, and, through you, to convey to the citizens of New York, the expression of their heartfelt thanks for this most gratifying address, as well as for the elegant and valuable testimonial which accompanies it.

It will be my pleasing duty to transmit to Captain Preedy this address and this beautiful token of your respect and regard, which I am sure will be cherished with feelings of pride and pleasure by him and by his children after him, as a gratifying proof of the manner in which the freemen of the city of New York appreciate and honor faithful efforts in the performance of high duty.

The Singing Societies then sang another ode composed by Mrs. Stevens, the words of which were as follows:

### ALL HAIL!

Air—*God Save the Queen.*

All hail across the main!
Thought thrills our Cable chain,
Hear, nations, hear!
Mind is victorious,
Columbia's made glorious,
While God watched over us;
Hear, nations, hear!

No storms the chain shall break,
Nations our greetings take;
Hear now our call!
Peace speaks from shore to shore,
Good-will be evermore,
While we this work adore:
Praise God for all!

After which Mr. Charles H. Haswell, President of the Board of Councilmen, advanced amid much applause, and, addressing Captain Hudson of the Niagara, said:

SIR—The Common Council of New York desire to express, through me, their Chairman, and on behalf of the city, the high appreciation of the efforts of yourself and of the officers and men under your command, in laying the Atlantic Telegraph Cable. It needed not this testimonial to make the people of this city, or of the nation, aware of the debt of gratitude they owe both to you and them; for the eyes of the civilized world have been upon you, and every step in your arduous course has been chronicled in the hearts as well as the memory of all.

The great originator and chief actor in this wonderful enterprise has availed himself of every occasion to say, that, but for you and those who served under you, all his efforts would have been fruitless.

Our recognition of those efforts can add nothing to their merit or increase the honors with which you are already crowned. Still, this metropolis, so deeply interested in the undertaking, cannot deny itself the pleasure of expressing its appreciation of your services and of offering, in common with the whole nation, its heartfelt congratulations on your success. To you, who have so long traversed the ocean, and hence know to what innumerable hazards such an enterprise was exposed to, its successful termination must appear more marvelous the oftener it is contemplated. When you linked your ship to the Agamemnon, and saw the flag of the latter disappear below the eastern horizon, the deep anxiety and sense of responsibility that weighed upon your heart we may attempt to imagine, but can never know. How well, how faithfully, how nobly, you and those under you, met that responsibility is now a matter of history. And never, sir, we sincerely believe, during the many years you have carried our national flag over the seas, have you greeted with such joy and exultation the shores of your native land as you then did the rugged heights of Newfoundland. The lapse of time will only increase the marvel as we contemplate that slender thread disappearing from sight in Valentia Bay, climbing down the lofty cliffs that form the eastern shore—now losing itself in the oozy bed of the lower abyss, and now rising over the swelling heights that intervened—still stretching on through the stirless solitudes of the deep, until finally it emerges into view in the wilderness of the western continent, the channel of instantaneous communication between the two worlds.

Of the vast results of this enterprise to the city and to the nation, I will not venture to speak. Of its value to the civilized world, it is impossible to conjecture. Columbus, when, through the dim twilight of morning, he descried the shores of a new world, did not dream with what destinies his name would be linked. The young surveyor, Washington, as he stood on the summit of the Alleghanies, and looked off on the forests stretching away on either side of him, could not conceive of the mighty nation that would one day call him "Father." So, no man can now foresee with what events, with what strange histories of men, your name and that of Mr. Field and others, who have taken part in this great work, will yet be joined; for you have inaugurated a system that will progress until a net-work of electrical nerves shall enwrap the world and a single central pulsation shall thrill the entire globe on which we dwell.

When that consummation shall be reached the changes that shall take place in the destinies of the race, the future developments of the great plans of Divine Providence, alone can reveal.

Sir, it is often observed that many of the great events in human history are connected with singular phenomena of nature. If they are wanting in this instance, we behold something quite as remarkable and far more gratifying. Is it not a strange coincidence that the two distinguishing features in the natural scenery of this State, Niagara and the Hudson, should be forever joined to this enterprise? While Niagara shall continue to thunder on our western frontier will the noble vessel Niagara be remembered by men, and so long as the Hudson shall roll its massive waters along our eastern frontier will the gallant commander of that vessel be enshrined in the hearts of the people of this nation.

In conclusion, sir, allow me to present to you, in behalf of the Common Council of the city of New York, this gold box, as a slight testimonial of the high value placed upon your services and as a mark of the esteem and respect in which you are and ever will be held.

Captain Hudson received the gift amid great cheering, and replied substantially as follows:

*Mr. Mayor and Gentlemen:*

I can hardly find words in which to express my deep sense of the honor you have this day conferred upon me, as the chief officer of the Niagara; and I feel the more unfit to make adequate acknowledgment of your kindness when I reflect that whatever share I may have had in the work, whose happy issue you now celebrate, it was the simple performance of a duty. We claim to have done nothing more. It is true our labor was tedious and engrossing, but of that we never thought. We had a high object before us, and the attainment of that was our first wish and is now our best reward. Whilst acknowledging our gratitude for your kindness, we feel that to others, as well as ourselves, equal praise is due. The British officers who shared in the work are deserving of the highest encomiums, and we can never forget how much is due to their assistance and how much we owe their country for her kindness. In conclusion, whilst repeating that we have done no more than our duty in the successful consumation of this great enterprise, we must not forget that our success is attributable not to us but to that Almighty and glorious Being whose creatures we are.

We had evidences in and around us, while we were at the work, that we had God with us, and I hope we will say on all occasions that "Unto Him and not unto us be all the praise."

Alderman McSpedon, as Chairman of the Committee of Arrangements, then presented the freedom of the city, with a gold medal, to Captain Dayman, of the Gorgon, addressing him as follows:

*Captain Dayman, Royal Navy;*

Sir—At the request of the Mayor and Corporation of the city of New York, I fulfill the duty, to me a most agreeable one, of presenting to you the accompanying Gold Medal, which has been struck to commemorate the great achievement in which your watchful care and zealous co-operation have borne so conspicuous a part. I pray you to receive it as a proof of the high sense which is entertained of your services by the inhabitants of this great city, the commercial metropolis of the United States, of which the Corporation is the official representative and organ, and to believe that on no occasion has this city, so notoriously the rewarder of merits and the promoter of international good-will, discharged the duties of this its high position with a livelier satisfaction or a more thorough conviction that its grateful acknowledgments have been well bestowed.

It would, sir, have afforded much pleasure to the citizens of New York, had it been possible for them to be honored by the company of your gallant brother officers, Captain Aldham, of Her Majesty's ship Valorous, and Captain Otter, of Her Majesty's ship Porcupine ; but, as circumstances have prevented their presence, I venture to request that you will receive this medal, similar to your own, for the former of those officers, and that you will convey to the latter the cordial thanks of this city, together with the assurance of its lively appreciation of his merits.

In conclusion, sir, I would be permitted to hope that when, in after years, you shall have risen to the highest ranks of your noble profession, you will occasionally indulge in a pleasant recollection of your reception to-day by the citizens of New York, whose best wishes for your future success will ever accompany you on your path to honor and renown.

Captain Dayman briefly replied, saying: He could not give adequate expression to the thanks which the kindness displayed on this occasion demanded, and hoped that the connection now established between England and America would be a bond of perpetual amity and good-will. For himself and his associates, who had had but an humble part in the laying of the Cable, he could only say that they felt doubly rewarded for all their anxiety and labor; rewarded by the consciousness of having performed their duty, and rewarded by this public recognition of that performance, a recognition for which they would ever be grateful.

Councilman Henry Arcularius made the presentation to Mr. Wm. E. Everett, Chief Engineer. In doing so, he said:

*Mr. Everett:*

SIR—We have met on one of the most interesting occasions—one in which, throughout our entire country and beyond the broad Atlantic, a common interest and gratification are felt.

Acting on the principle that honor should be acknowledged where due, the municipal authorities of New York have determined to convey to you in these public circumstances, and before this vast throng of friends and citizens, by suitable testimonial, their deep sense of public obligations to you for the part you have most ably and succesfully taken in laying the Ocean Telegraphic Cable.

I esteem it a special honor to be the medium of communicating their views and feelings, and am satisfied that I speak not for them only, nor for the thousands now before me, nor for the one and a half millions of our city, and the cities and country adjacent, but for the twenty-seven millions of our country, and for the friends of science and humanity everywhere, when I congratulate you on the accomplishment of your great undertaking.

Sir. The half century just completed has been so crowded with memorable events, and as memorable characters, that it has a distinction far beyond any similar period in our world's history. The mind is filled with admiration in tracing the progress of science, its discoveries, its triumphs, and its successful adaptations to the great purposes of life.

No less remarkable has been the advance in the arts, both æsthetic and useful—and in the developments of the resources and hidden wealth of the earth.

Religion has kept pace in the onward and upward march and marked this as an age of beneficent aggression and a world-wide charity, and has seen glorious triumphs over the darkness and degradation of paganism in the islands of the Pacific as well as in Asia and Africa.

To live in this period, so rife in great and glorious events, is a high privilege. The public mind has been trained to the expectation of great things, and they come in rapid succession.

To you and your coadjutors is due the honor of ushering in the second half of the century by an event of crowning excellence. It is THE WORK OF THE AGE, important beyond any event in science or art of ancient or modern times, and following beautifully in sequence to the widely introduced invention of our countryman, Professor Morse. We may well ask what class of men, what departments of busy life, what interests of literature, or science, or religion, what nation, however remote, what point of time in the future of the world, are not to be directly or indirectly affected by it? We are lost in the calculation of its results to the present and the future! It seems to us, to be *the benefaction of science to the world and all ages of the world.*

When we look at it in the aspect of a tie binding America and England, (the children to their sires), in the first instance, and then as the great movement which will lead to the union of nations in metallic bonds, and these receiving their strength more from the feelings of a universal brotherhood than their material, we cannot but specially thank the Divine Providence which has so favored it. It is designed clearly to be *the great peace agency,* more potent than any or all the organizations which have been devised.

As with men in smaller communities, so with nations, the more the facilities of intercourse are multiplied, the more perfect the means of understanding each other's views and policy, the fewer will be the sources of disquiet and difficulty, the more ready will be the mutual accommodations to that which will promote mutual benefit, and hence the more certain the prevalence of an intelligent peace policy—which is the true policy of our higher civilization. The true benefactor is he who contributes most to secure this end, and this honor have you and all the gentlemen connected in the work and the Governments which have favored it.

At the time we have been intensely occupied with

your successful efforts in uniting two worlds, have co-curred scenes of almost unparalleled splendor on the coast of France, in commemoration of the completion of vast arrangements for national defense and preparation for war. Two of the most distinguished crowned heads of the world have united in doing honor to the occasion. All was on a scale of magnificence worthy those imperial personages, and was finished in a blaze of glory. We admire the pageant from our distant position, and would have been enraptured with the ocular view. But for what was all this outlay of royal resources—this engineering skill, and this series of imposing fêtes—simply, on the completion of a work of national intimidation; whose associations, if it does not become moss-grown, (which God grant it may), will become historic with scenes of bloody war, and devastation, and the sacrifice of human life, and the multiplication of human sorrows. It is a matter of high satisfaction that your work has to do with the interests of a higher civilization, with the advancement of human good, and will render comparatively nugatory the Cherbourgs and Sebastopols of the world.

A foreign authority truly says: "The Atlantic Telegraph will exercise an influence more important and beneficial than the armies and fleets of a hundred kingdoms."

This day it is proclaimed that the inhabitants of the Old and of the New World have entered upon a new era.

In the progress of your efforts you have encountered difficulties and disappointments, some anticipated and others unforeseen; and the manner in which you have risen above all, and, with steady aim and true courage, persevered to the happy issue, has won universal admiration.

We watched your progress with deep solicitude and quickened sympathy; the delay of tidings for a succession of days awakened mingled hopes and fears. The all-absorbing daily inquiry was, what tidings of the Telegraph Fleet? When the word came, 'The Cable is laid,'—it was the only news men listened to—it passed from man to man, and the very children caught the strain.—'The Cable is laid'—over our country it sped on lightning's wings; hands were lifted in amazement, eyes filled with the tears of joyful emotions, and thankful returns were rendered to the benignant Providence which had crowned the enterprise. All seemed to feel the sublime sentiment of Captain Hudson: 'God has been with us; the Telegraph Cable is laid without accident, and to Him be all the glory.'

Sir, we repeat our congratulations; we welcome you to the honors of this day, not as the soldier returned from war, with sword bathed in blood—(yet this would we do in a righteous cause),—but as the representative of the true spirit of our age, the spirit of beneficent progress—the spirit of advancing civilization—the spirit of "peace on earth and good-will to men."

Every agent in this work has a share in the honorable results; yet the more controlling minds may well receive the higher meed. Sir, yours was the special responsibility, as the work belonged specially to your department. You will have many imitators. This is the law of all successful experiments and inventions; but to your Company will belong the honor of laying the *first Atlantic Telegraph;* and this will be the highly-prized heir-loom to your children and friends, and if you accomplish no more, this will be honor enough for a lifetime.

And now, gentlemen of the Common Council, we may be permitted to tender particular congratulations to Assistant Engineer Woodhouse. His position of responsibility was sustained in the spirit of that high reputation he had already won on the Baltic and on the Irish sea His country knows how to appreciate and reward talents and character like his. We bespeak for him a chief place among the engineers of glorious old England; he has already achieved it.

Chief Engineer Everett belongs to us—he was "to the manor born;" the Empire State claims him as her son, and enrols his name from this day on the catalogue of her worthies. We admire his frank, unsought submission of plans for adapting his ship for reception of the Cable. We admire the magnanimity of the British Governors of the Company in soliciting from our Government that he might be furloughed to take the chief-engineering charge. We admire his success in combining and setting up on board the two frigates the apparatus for "paying out," on which everything depended. We admire his stern perseverance, and his sacrifice of personal comfort and repose to the success of the enterprise. We bespeak for him from our country a position becoming such merit.

May they all be spared to see for many years the benefits resulting from this work, and may they enjoy the constant smiles of a benignant Providence.

With this we unite in the noble sentiment of the President's response to England's Queen—alike worthy of himself and the people over whom he is placed:

"May the Atlantic Telegraph, under the blessing of Heaven, prove to be a bond of perpetual peace and friendship between the kindred nations, and an instrument destined by Divine Providence to diffuse religion, civilization, liberty and law, throughout the world.

"In this view, will not all nations of christendom spontaneously unite in the declaration that it shall be forever neutral, and that its communications shall be held sacred in passing to their places of destination even in the midst of hostilities."

Mr. Everett, in acknowledgment, replied as follows:

*Mr. Mayor, and Gentlemen of the Common Council of the city of New York:*

The honor which you have conferred on the Chief Engineers of the Atlantic Telegraph Cable, is one which might naturally be expected from your generous enthusiasm and intelligent appreciation of every movement calculated to promote the interests of the human race. For myself, I feel deeply the compliment you have tendered, and shall keep it in lasting remembrance, as coming from the first city of the United States, the land of my birth and my affections. On behalf of my associates, I beg to offer you my thanks, and am confident that I can but ill express their sense of your kindness. These tokens of your good will we shall ever preserve, not only as souvenirs of this great event which you celebrate to-day, but as constant reminders that "the path of duty is the path to fame."

Gentlemen, on behalf of myself and my associates from the other side of the water, I return you our warmest acknowledgments.

Alderman James Davis then came forward, and presented to Mr. Peter Cooper, as the representative of the New York, Newfoundland and London Telegraph Company, and the Atlantic Telegraph Company, the following address:

*Gentlemen of the New York, Newfoundland and London, and Atlantic Telegraph Companies:*

We, the Common Council of the city of New York, in our official capacity, and on behalf of our constituents, as well as of millions of our fellow-citizens, beg leave to congratulate you on the successful issue and glorious accomplishment of the great work in which you have been engaged. The Atlantic Telegraph Cable, connecting the thoughts of America with those of Europe by instantaneous communication, has been successfully laid.

This could not have been done, as it has been, without your co-operation. Science is only a pity, when unsupported by pecuniary means to make it practical. These means you have liberally supplied.

The Genoese sailor, Columbus, was as scientific *before* the discovery of America as *after*. Science had taught him to believe that there was, or ought to be, a continent away off in the Western Ocean, and enthusiasm, scarcely less important than science itself in great novel enterprises, almost persuaded him that, if his continent did not appear exactly where he hoped to find it, he could pluck it up to the surface from the depths of the sea. But even Columbus would have languished into an obscure grave if the great Isabella, with the enlightened generosity of a noble queen, and in disregard of the supposed wisdom of her own cabinet, had not flung her private jewels into the scale of the Italian mariner.

What Isabella did for Columbus, and indeed for the whole human race, you, gentlemen of the New York, Newfoundland and London, and Atlantic Telegraph Companies, have done for the projectors of the Submarine Cable which now unites Europe and America. Their science, too, would have been a special pity, if you had not stood up and indorsed it with your approval, your reputation, and your fortunes.

It is your merit to have enlisted the Governments of the United States and of Great Britain in your noble cause—and we rejoice that they have so magnanimously co-operated with you in the successful execution of a work not conceived in any narrow view of mere national policy, but intended for the benefit of the entire human race. These Governments sent, we may say, the pride of their navies to aid you; and from the gallant commander on the quarter deck down to the humblest sailor before the mast, they placed their science, their authority, their prompt obedience to orders, and above all, their fervent sympathy, at your disposal. Thanks to them all from the Common Council of the city of New York, the expression of which we take the liberty of requesting you to communicate as you shall think proper.

We should not fulfill the pleasing duty which public sentiment imposes on us, if we omitted to mention the special liberality of the several Colonial Governments through whose territory, on this side of the Atlantic, your electric wire is to pass. The encouragement given by the

Legislature of Newfoundland is particularly worthy both of praise and of imitation. It was magnanimous and politically wise withal. Our own North-eastern sister, Maine, also behaved in a style of which her own citizens, and her sister States throughout the Union, may well be proud.

And now, gentlemen of the New York, Newfoundland and London, and Atlantic Telegraph Companies, what shall we say of the results of your enterprise, as they shall develop themselves in all time to come?

On this part of the subject it becomes us to speak with modesty and diffidence. It involves a problem which future experience can alone solve. Nevertheless, in contemplating he prospective results of your labors, we anticipate great benefits to the human family at large.

Christianity is the basis of civilization. The nations that have rejected it from the beginning remain in *statu quo.* They have not advanced. Other nations that have expelled after having once received it, have invariably relapsed intô their primitive stupidity. The electric Cable may be made, in a certain sense, the harbinger of Christianity to all peoples. It never can be the messenger proper. But it will announce or insinuate in a thousand ways the truth of revelation, in lands and amid communities where the living missionary dare not, in the first instance, utter the same, except at the immediate peril of his life.

If civilization follows in the footsteps of Christianity, who shall venture to limit or circumscribe the beneficent results of the work which you have sustained through all the vicissitudes of its history?

The thought of America will blend itself instantaneously with the thought of Europe, Asia and Africa; and America will receive theirs in prompt return.

Almighty God, the Creator of the Universe, made this earth of ours in the form of a sphere, not a plane. If it had been a plane, the operation of sin, working through the passions of nations and individuals, would have driven the weak to the extreme brink of humanity and hurled them over the precipice, once for all. But our merciful Father has appointed it otherwise. Our planet is spherical. It is cut up on its surface by oceans, seas, mountains, rivers—it is diversified by latitudes and longitudes—it is peopled by all tribes, tongues and nations, and yet its inhabitants have been perpetually struggling to overcome these physical boundaries, and to hug each other more and more, as old acquaintances of the same original and universal family too long estranged and separated.

If such are the natural tendencies of enlightened humanity, we have still greater reason, gentlemen, to congratulate you on the great work which you have aided and sustained to a triumphant accomplishment. Go on! Your Cable is extended under the stormiest ocean that is found on the surface of our globe. You have banded together, by your Cable, two of the most enlightened nations that can be found on the same surface. Your work is, perhaps, done. But you can influence both the present generation and posterity toward its extension through all present space and future time.

It is supposed that every Cabinet of human government has a special department which is not honest, but which comes under the head of "Intriguing Diplomacy." Your

Cable will put an end to this special department, and it is to be hoped that no cabinet minister, no private individual, no missionary, no merchant, on either side of the Atlantic, will be so profane as to tax Heaven's lightning, which is to abide in your Cable, with the message of a lie.

We do not enter, gentlemen, into the relations which your noble work is to sustain toward the commercial life of great communities on this side of the ocean or the other. These relations cannot but be improved by the facilities of inter-communication which your enterprise has provided for them.

Neither can it fail to be the means of promoting peace on earth. It presents, indeed, for abuse as well as use, the mysterious power and passive obedience which are its own peculiar properties. But to both continents it preaches silently the doctrine of peace. No tiny infant ever slept more tranquilly on its mother's breast than your tiny Cable reposes on the bottom of the ocean. It rests on the solid land, and has the deep blue broad seas for its covering. The tempests, in their delirium of fury on the water's surface, which tried its strength, and well-nigh deprived the world of its services for a time, these shall disturb it no more. The tranquillity of its repose will be almost sufficient to forbid the transmission of angry messages, or, if transmitted, it will mollify the asperity of temper in which they were dictated. And thus it will silently suggest to thinking men that they should rise above the moral atmosphere in which human passions exercise their violent and destructive influence.

In fine, gentlemen, the Common Council of the city of New York, in their own name, as before expressed, and in the name of their fellow-citizens, with a full heart, beg leave to congratulate you, again and again, on the successful termination of your glorious enterprise.

To which Mr. Cooper replied, as follows:

It is, sir, with extreme hesitation that I, as President of the New York, Newfoundland and London Telegraph Company, venture to reply to remarks so eloquently expressed in honor of the success that has crowned the united efforts of the New York, Newfoundland and London, and the Atlantic Telegraph Companies. That success, sir, has been secured by a long-continued course of persevering labor, under circumstances the most discouraging; labors that required the indomitable courage, the farseeing and electrifying mind of Cyrus W. Field to inspire and stimulate.

This work, now accomplished, has caused more hearts to beat in joyful hope of a brighter and better day for suffering humanity throughout the world, than was ever before inspired by any previous triumph of mind over matter. Such a triumph is calculated to elevate our race to a perception of the true dignity and responsibility of their mission, being clothed, as they are, with power to keep the great garden of the world without and the world within, and thus to subdue and hold dominion over all the earth and every living thing. This glorious power bequeathed to man will, I trust, cause knowledge to cover the earth, as the mighty waters cover the umbilical cord that now binds the mother Continent to her child. Through this vital cord a new regenerating and invigorating life will flow on and ever from the world's heart, through arteries, veins and nerves of iron, into every fibre of the great body of humanity. This life, this power, this knowledge, cannot fail finally to bring man into sympathy with the sorrows and sufferings of his brother man throughout the world. This mighty power to communicate knowledge will finally bring life and immortality to light in the intellectual heart of man. It will, with its electric power, spread over the world the glad tidings of that great joy, which shall be to all people. I rejoice in the hope that the time will come when they learn war no more, when they will beat their swords into ploughshares and their spears into pruning-hooks. Then the light of science will shine on the unfolding and ever-brightening leaves of creation with an all-animating power—a power that will awaken the slumbering energies of man to aspire after all that is true, beautiful and good. Encouraged by what has already been accomplished—having sent the electric fire through an almost fathomless ocean—carrying with it expressions of sympathy flowing from the very heart of a kindred people, and then giving to the world the glorious news of peace restored to a region where war, with its desolating power, has spread its terrors over more than one quarter of the great family of man; after such a triumph of mind over matter, what limits can we set to what is possible to be accomplished? I am sure, sir, that every man, woman and child within sound of my voice, will join me in heartfelt thanks to the enlightened Governments that have lent their aid to enable us to achieve the victory we this day celebrate. In conclusion, sir, I will say, in behalf of the companies which I in part have the honor to represent, that language fails to express the admiration we feel for all the officers and their devoted sailors, and the electricians and engineers, by whose united labor and skill a glorious treasure has been secured to the world.

When the applause, with which Mr. Cooper's response was received, had died away,

Mr. David Dudley Field, the orator of the occasion, was introduced amid renewed plaudits, and said:

Ladies and Gentlemen—When Morse discovered the applicability of electricity to the communication of intelligence, it might have been foreseen that the limits of the application were to be measured only by the power of stretching the electric wire, and of transmitting through it the electric current. It occurred, no doubt, to different minds that the telegraph would one day be carried across the ocean and around the globe; and, for aught I know, plans may have been formed for doing the work. I have been requested to give you the history—a condensed epitome it must be—of the first success—the first attempt, and, I might add, the first practicable plan in the development of this great idea of an Ocean Telegraph. My connection with the undertaking from its commencement—my position as counsel for those who have done the most to carry it through—have made it appear to others fitting that I should perform this service. In its performance I trust that I shall say nothing unbefiting my personal relations to any of the actors. I am not here to praise, but to relate.

Two years previous to 1854, there had been incorporated by the Legislature of Newfoundland, a Company by the name of the Newfoundland Electric Telegraph Company, the purpose of which was to connect by telegraph that island with the main land of America. A telegraph across the ocean was not a part of the scheme. It contemplated a connection with Europe by means of steamers plying between Newfoundland and Ireland.

This Company proceeded a little way and failed, leaving a debt of some fifty thousand dollars, due chiefly to laborers. In this emergency, and some time in February, 1854, Mr. Horace B. Tebbetts and Mr. Frederick N. Gisborne, officers then of that Company, applied to Mr. Matthew D. Field to help raise additional funds by a sale of bonds or stock. The gentleman thus applied to came to Mr. Cyrus W. Field and myself. We had several conversations together on the subject. Then it was, that the thought of extending the line across the Atlantic suggested itself. Mr. Cyrus W. Field wrote to Lieutenant Maury to inquire about the practicability of submerging a Cable, and consulted Professor Morse about the possibility of telegraphing through it. Their answers were favorable.

On receiving them, it was agreed between Mr. Cyrus W. Field and myself, that as nothing could be done under the charter of the Newfoundland Electric Company, we would endeavor to form a new Company, to take a surrender of the charter of the former Company, purchase its property, pay its debts, and obtain another charter to effect a direct telegraphic communication with Europe. The first step was to procure the co-operation of a few persons whose character and resources would be a guaranty that the work had been undertaken in earnest. Four men were invited, whose names you all know—Peter Cooper, Moses Taylor, Marshall O. Roberts, and Chandler White. They met Mr. Cyrus W. Field and myself at his house, where, around a table covered with maps, plans and estimates, the subject was discussed for four successive evenings, the practicability of the undertaking examined, its advantages, its cost, and the means of its accomplishment. The result of the conference was the agreement of all the six gentlemen to enter upon the undertaking. Mr. Cyrus W. Field, Mr. White and myself, were to proceed to Newfoundland to procure a charter and such aid in money and privileges as the Government of that island could be induced to give. The agreement with the Electric Telegraph Company and the formal surrender of its charter, were signed on the 10th of March, and on the 14th we left New York, accompanied by Mr. Gisborne. The next morning we took the steamer at Boston for Halifax, and thence, on the night of the 18th, departed in the little steamer Merlin for St. John's, Newfoundland.

Three more disagreeable days voyagers scarcely ever passed than we spent in that smallest of steamers. It seemed as if all the storms of winter had been reserved for the first month of spring. A frost-bound coast, an icy sea, rain, hail, snow and tempest, were the greeting of the telegraph adventurers in their first movement towards Europe. In the darkest night, through which no man could see the ship's length, with snow filling the air and flying into the eyes of the sailors, with ice in the water, and a heavy sea rolling and moaning about us, the Captain felt his way around Cape Race, with his lead, as the blind man feels his way with his staff, but as confidently and as safely as if the sky had been clear and the sea calm, and the light of morning dawned upon deck and mast and spar, coated with glittering ice; but floating securely between the mountain gates of the harbor of St. John's.

In that busy and hospitable town the first person to

whom we were introduced was Mr. Edward M. Archibald, then Attorney-General of the colony, and now British Consul in New York. He entered warmly into our views; and from that day to this has been an efficient and consistent supporter of the undertaking. By him we were introduced to the Governor (Kerr Bailey Hamilton), who also took an earnest interest in our plans. He convoked the Council to receive us and hear an explanation of our views and wishes. In a few hours after the conference the answer of the Governor and Council was received, consenting to recommend to the Assembly a guaranty of the interest of £50,000 of bonds, an immediate grant of fifty square miles of land, a further grant to the same extent on the completion of the telegraph across the ocean, and a payment of £5,000 towards the construction of a bridle-path across the island, along the line of the land telegraph.

Mr. Cyrus W. Field thereupon, on the 25th of March, took the return steamer from St. John's, on his way to New York, in order to fit out a steamer for the service of the Company, while his two associates remained in Newfoundland to obtain the charter and carry out the arrangements with the former Company. They continued there nearly five weeks, during which, after many discussions and negotiations, the charter was at length obtained, and the $50,000 of debt of the old Company were thereupon paid.

The charter was liberal and provident. After declaring that it was "advisable to establish a line of telegraphic communication between America and Europe, by way of Newfoundland," it incorporated the associates for fifty years, established perfect equality in respect to corporators and officers between citizens of the United States and British subjects, allowed the meetings of the stockholders and Directors to be held in New York, or in Newfoundland, or in London, conceded the exclusive right to establish a telegraph from the continent of America to Newfoundland, and from Newfoundland across the ocean, granted fifty square miles of land ; and further provided that " so soon as the said Company shall have actually established a communication across the Atlantic Ocean, by means of a submarine Cable or wire from this island, the said Company shall receive from the Government of this island a grant of fifty square miles of ungranted and unoccupied wilderness land, to be selected by the said Company, in addition to the grants hereinbefore mentioned," a provision subsequently extended so as to permit the Company to establish the communication by an auxiliary or associate Company.

It were long to tell how the Government and people of Newfoundland nurtured this enterprise in its commencement—how they have stood by it, through its various fortunes, till its triumphant consummation. That vast island, projected into the North Atlantic, lifting above the sea its cliffs of everlasting and immovable rock, beckoning, as it were, to Europe, seems framed by Providence for one of the pillars of that Cable which is to bind the continents together. Its broad interior, baffling the explorer, its cold and gloomy morasses, its dark and frowning headlands, its deep and tranquil bays, and harbors innumerable, take not such hold of the imagination as its support of that wondrous line which, lost forever to human eyes, is to be the highway of thought between the Old World and the New.

Take the map, and see where the civilized portions of

the two hemispheres approach nearest to each other : two islands stand there face to face. The highlands of Trinity answer to the highlands of Valentia. Between them rolls the stormiest sea of all the world save one. It is the gateway through which pass the icebergs from the Pole. Once a year, and sometimes for forty days together, a continuous field of ice moves down from the north at the rate of two or three miles the hour. But far beneath there is tranquil water and an even surface. The plummet has sounded all that sea, and found, at an average depth of about two miles, a nearly level bottom covered with the smallest sea shells, which must have been deposited in the lapse of ages and fallen through the still water as the snow falls through the still air.

In the early part of May the two gentlemen who had remained behind, in Newfoundland, rejoined their associates in New York, and there the charter was formally accepted and the Company organized. As all the associates had not arrived till Saturday evening, the 6th of May, and as one of them was to leave town on the morning of Monday, it was agreed that we should meet for organization at six o'clock of that day. At that hour they came to my house, and, as the first rays of the morning sun streamed into the windows, the formal organization took place. The charter was accepted, the stock subscribed and the officers chosen. Mr. Cooper, Mr. Taylor, Mr. Field, Mr. Roberts and Mr. White, were the first Directors. Mr. Cooper was chosen President, Mr. White Vice-President and Mr. Taylor Treasurer. Thus was inaugurated that great enterprise whose completion we celebrate to-day. The plan was formed, the arrangements made, and the work begun. What followed was the execution of the great design.

From the 8th of May, 1854, to the 5th of August, 1858, there passed scarcely four years and three months; but they were as fruitful of anxiety and toil as of successful results. The land line across the island of Newfoundland—upwards of four hundred miles—was first to be made. This was a work of incredible labor. The country was for the most part a wilderness of rock and morass, "a good and traversable bridle road eight feet wide," with bridges of the same width, had to be made the whole distance; materials and provisions had to be transported first from St. John's to the heads of the different bays on the southern coast, and afterwards chiefly on men's backs to the line of road. The first year Mr. White, as Vice-President, directed in person the operations; the second and third year superintendents were sent down. In addition to the land line in Newfoundland another of one hundred and forty miles in Cape Breton was constructed, and contracts made with companies in Nova Scotia, New Brunswick, Maine, New Hampshire, Massachusetts, Connecticut and New York, to connect their lines with the Newfoundland line. Then there was the submarine line between Newfoundland and Cape Breton, eighty-five miles in length, and another thirteen miles long, across Northumberland Straits, to Prince Edward Island. To procure these Mr. Cyrus W. Field visited England twice—once in December, 1854, and again in January, 1856. The first attempt to lay the submarine line across the Gulf of St. Lawrence was made in 1855, and was unsuccessful. A second attempt, made the next year, succeeded. Thus was completed the chain of telegraph from New York to the eastern coast of Newfoundland, and the projectors now stood upon the shore of the Atlantic in their progress eastward.

The whole expense thus far, with very trifling exceptions, had fallen upon them—Mr. Cyrus W. Field having made the largest contributions—amounting to more than two hundred thousand dollars in money—and Mr. Cooper, Mr. Taylor, and Mr. Roberts, each a little less. No other contributors beyond the six original subscribers, had come in except Professor Morse, Mr. Robert W. Lowber, Mr. Wilson G. Hunt and Mr. John W. Brett. The list of Directors and officers remains to this day as it was at first, except that Mr. Hunt, as Director, has taken the place of Mr. White, who died in 1856, and that Mr. Field is Vice-President, and Mr. Lowber Secretary. In all the operations of the Company, thus far, the various negotiations, the plan of the work, the oversight of its execution, and the correspondence with the officers and others mainly devolved upon Mr. Cyrus W. Field.

The greatest and most difficult part of the original design still remained to be executed, and that was the submarine Cable from Newfoundland to Ireland. The distance was one thousand nine hundred and fifty statute miles ; the sea was stormy and uncertain ; no submarine line of more than three hundred miles had then been attempted. In anticipation of the task now to be undertaken, Mr. Field, on his first visit to England, in 1854, had invited manufacturers to furnish him with specimens of cable which they would recommend and estimates of its cost, and he had entered into correspondence with various persons on the subject. In 1856 he procured an order from our Government, under which Lieutenant Berryman made soundings of the Atlantic, between Newfoundland and Ireland. Lieutenant Berryman sailed on that service on the 18th of July, and the next day Mr. Field sailed for England, having received the formal authority of the Company to make arrangements in England for the submarine line, either by a subscription to this Company or by organizing a new Company as auxiliary or associated with this. In England he had invited the co-operation of Mr. Brett, a gentleman of great experience, who, in 1851, formed a Company which had laid the first submarine cable from England to France. He afterwards brought in Mr. Edward O. W. Whitehouse, electrician, and Mr. Charles T. Bright, engineer—both gentlemen, of high scientific attainments. These four gentlemen on the 29th of September, 1856, entered into a formal agreement to use their exertions for the formation of a new Company, to be called the Atlantic Telegraph Company, the object of which should be "to continue the existing line of the New York, Newfoundland and London Telegraph Company to Ireland, by making, or causing to be made, a submarine Telegraph Cable for the Atlantic." This done, Mr. Field issued, on the 1st of November, 1856, a circular signed by him, as Vice-President of the New York, Newfoundland and London Telegraph Company, from which I cannot forbear making the following extracts:

"In April, 1854, a Company was incorporated by act of the Colonial Legislature of Newfoundland, for the purpose of establishing a line of telegraphic communication between America and Europe. That Government evinced the warmest interest in the undertaking, and in order to mark substantially their sense of its importance and their desire to give to it all the aid and encouragement in their power, they conferred upon it, in addition to important privileges of grants of land and subsidy, the sole and exclusive right of landing a telegraphic line on the shores within their jurisdiction, comprising, in addition to those of Newfoundland, the whole Atlantic coast of Labrador, from the entrance of Hudson Straits to the Straits of Belle Isle.

"This act of the Colonial Legislature was subsequently ratified and confirmed by her Majesty's Government at home. The Company also obtained in May, 1854, an exclusive charter from the Government of Prince Edward Island, and afterwards from the State of Maine, and a charter for telegraphic operations in Canada.

"The exclusive rights absolutely necessary for the encouragement of an undertaking of this nature, having thus been secured along the only seaboard eligible for the western terminus of an European and American Cable, the Company in the first instance commenced operations by proceeding to connect St. John's, Newfoundland, with the widely ramified telegraph system of the British North American provinces and the United States.

"This has been recently completed by the submersion of two Cables in connection with their land lines; one, eighty-five miles in length, under the waters of the Gulf of St. Lawrence, from Cape Ray Cove, Newfoundland, to Ashpee Bay, Cape Breton; the other, of thirteen miles across the Straits of Northumberland, connecting Prince Edward Island with New Brunswick. Electric communication is thus established direct from Newfoundland to all the British American colonies and the United States. On the Irish side lines of telegraph have been for some time in operation throughout the country, and are connected with England and the Continent by submarine cables. The only remaining link in this electric chain required to connect the two hemispheres, by telegraph, is the Atlantic Cable. The New York, Newfoundland and London Telegraph Company being desirous that this great undertaking should be established on a broad and national basis, uniting the interests of the telegraph world on both sides of the Atlantic, have entered into alliance with persons of importance and influence in the telegraphic affairs of Great Britain; and, in order at the same time to obtain the fullest possible information before entering upon the crowning effort of their labors, they have endeavored to concentrate upon the various departments of the undertaking the energies of men of the highest acknowledged standing in their profession, and of others eminently fitted for the work, who were known to have devoted much time and attention to the subject."

After detailing the results of the investigations, the circular proceeded:

"All the points having a direct practical bearing on any part of the undertaking have thus been subjected to a close and rigid scrutiny, the result of this examination proving to be in every respect of the most favorable character. It remained only that those possessing the required power should take the initiative. The New York, Newfoundland and London Telegraph Company, possessing, in virtue of their charter, all the necessary powers, deputed their Vice-President to visit England in the summer of the present year, and they gave him full authority to make, on their behalf, such arrangements as should seem to him best fitted to carry forward the great work. The outline of the formation of the Atlantic Telegraph Company, which will be found in the appendix, will sufficiently explain the nature of these arrangements."

Without waiting for the formation of the new Company, Mr. Field, on behalf of the Newfoundland Company, made application to the British Government for its aid in ships and money, and received on the 20th of November a letter from the Treasury, which I am tempted to read, promising ships to assist in laying the Cable, and a fixed yearly sum in payment for Government messages. He also personally solicited bankers and merchants in London for subscriptions, and, with Mr. Brett, visited Liverpool and Manchester to address public meetings. He subscribed £100,000 towards the capital of £350,000, and Mr. Brett followed with a subscription of £25,000. A day or two after the Treasury letter was received, the subscriptions were closed, when it was found that the applications for stock exceeded the capital by about £30,000, so that on the final allotment Mr. Field had eighty-eight shares, of £1,000 each, and Mr. Brett twelve. To show the feeling which had been excited in England, it is worth mentioning that many persons subscribed for shares, not for profit, but that they might have a part in the undertaking, and among others Mr. Thackeray and Lady Byron.

Too much praise cannot be awarded to the English Government and people for the zeal with which they came forward in answer to the call made upon them. Money was obtained from individuals as freely as it was wanted, and the Government outran the people.

Returning then to America, Mr. Field, with his American associates, made application to the Government of the United States for aid similar to that given by the English Government, and he also applied to individuals for a participation with him in the stock he had taken. Congress voted the aid requested, after a vehement opposition, against which the measure was carried in the Senate by a majority of one. Of the stock twenty-seven shares were taken in the United States.

All things being now ready, the first attempt to lay the Cable was made, as you all know, in August, 1857. There had been assembled in the harbor of Valentia three ships of the English and two of the American navy. There was the Agamemnon, recent from the fires of the Crimean war; she had borne the flag of the English Admiral over the waters of the Euxine; she had now laid her armament aside, and taken the burden of half that coil, for the laying of which she will be hereafter more famous than if she had forced the harbor of Sebastopol. There was the Niagara, the largest ship of our navy, made for the heaviest cannon of naval warfare, her armor never yet put on, but laden instead with the American half of the precious burden. There were the two attending ships, the Leopard and the Susquehanna, and the Cyclops, surveying ship, just returned from the verification of Lieutenant Berryman's soundings. The Lord-Lieutenant of Ireland had come to wish them "God speed" in the name of his sovereign and her people Everything promised success, and as the great ships moved out of the harbor, the highlands of Valentia shone brightly in the morning sun; while behind them the grand old mountains about Killarney towering above th lakes—those miracles of beauty—appeared to smile and beckon the ships westward, for, to the excited imagination, it seemed as if the inanimate mass were conscious of the great act about to be performed, and looked impatient toward the west, which it had faced in silence since the world began, but to which it was soon to speak in tones inaudible to human ears, yet signifying the thoughts and wishes of men.

The expedition thus prosperously begun was, however, doomed to sudden disappointment; for, on the fourth day out, the Cable parted, and the ships made their way to England. The undertaking being thus suspended for the year, Mr. Field returned to America. He was sôon, however, recalled to England to assume the management of the enterprise. Arriving in that country on the 16th of January, he was, on the 27th of the same month, appointed the general manager—an appointment which he accepted without compensation; and, by a subsequent resolution, every person in the employment of the Company was placed under his control.

The precise share which each person had hitherto borne in the great undertaking is easily measured by the narrative which I have given. The Directors of the Company, in their report to the stockholders on the 18th of February last, thus state the share of one of them:

"The Directors cannot close their observations to the shareholders without leaving their warm and cordial testimony to the untiring zeal, talent and energy that have been displayed on behalf of this enterprise by Mr. Cyrus W. Field, of New York, to whom mainly belongs the honor of having practically developed the possibility, and of having brought together the material means for carrying out the great idea of connecting Europe and America by a submarine telegraph."

"He has crossed the Atlantic Ocean no less than six times since December, 1856, for the sole purpose of rendering most valuable aid to this undertaking. He has also visited the British North American colonies on several occasions, and obtained concessions and advantages that are highly appreciated by the Directors, and he has successfully supported the efforts of the Directors in obtaining an annual subsidy for twenty-five years from the Government of the United States of America, the grant of the use of their national ships in assisting to lay the Cable in 1857, and also to assist in the same service this year—and his constant and assiduous attention to everything that could contribute to the welfare of the Company, from its first formation, have materially contributed to promote many of its most necessary and important arrangements. He is now again in England, his energy and confidence in the undertaking entirely unabated; and at the earnest request of the Board he has consented to remain in that country for the purpose of affording to the Directors the benefit of his great experience and judgment, as general manager of the business of the Company connected with the next expedition.

"This arrangement will doubtless prove as pleasing to the shareholders as it is agreeable and satisfactory to the Directors."

Everything being now ready for the second trial, which it was determined to begin—not at the shore, but in mid-ocean—the squadron departed from Plymouth on the 10th of June. It consisted of the Agamemnon and Niagara, to lay the Cable, and the Valorous and Gorgon (both English,) as attendant ships, the Susquehanna being kept away by the yellow fever, which had broken out on board, and the Gorgon taking her place, while the Valorous took the place of the Leopard of the previous year. The officers and crews of these vessels were picked men. Captain Preedy of the Agamemnon, and Captain Hudson of the Niagara, are as accomplished and gallant commanders as ever trod the quarter-deck; and Captain Dayman in the Gorgon, and Captain Aldham in the Valorous, fitly represented the spirit and honor of the English navy. Then, what a company was there of engineers and electricians! I need only name Mr. Everett, to whose genius the paying-out machinery was due; Mr. Bright, Mr. Woodhouse, Mr. Canning and Professor W. Thompson, to show that everything was provided which science and experience could suggest. Stately ships, illustrious company, and a richer freight than ever filled the argosies of Spain when Spain was mistress of the Indies.

On the open sea they found not that calm weather which they had been led to expect, but violent storms. A hurricane saluted them on their approach to mid-ocean. They gained, however, on the 26th of June, the point desired, spliced the Cable and steered in opposite directions. The Cable parted after about five miles had been paid out. They returned and made another splice on the same day and started again. A second time the Cable parted, and about seventy miles more were lost. Nothing daunted, they returned and made a third splice. All went well until two hundred and sixty miles more had been laid in the sea, when another break occurred, and the ships, according to the preconcerted arrangement, returned to Queenstown.

Anxiously had they been expected at Valentia; from whose headlands eyes were strained every day to catch the first glimpse of the returning Agamemnon, rising out of the western horizon. I have it on good authority that the Queen was waiting for the signal to go herself and receive the Cable. Would it not have been an admirable sight to see that illustrious lady, the foremost woman of all the world, sovereign of so many lands, the heir of the kings of our forefathers, receiving from her gallant seamen that line which was to repair with material better than allegiance the broken chain which once bound together the Anglo-Saxon-Celtic races, in every quarter of the globe?

The ships being returned, the Directors were summoned to meet in London. This was the time to try the fortitude of men. It was the agony of the enterprise. If it had been abandoned then, who can tell when it would have been resumed? The meeting of the Directors took place on the 14th of July, and then the fate of the undertaking was decided. There were sixteen acting Directors; of these, six were absent; another, the Vice-Chairman, was so dissatisfied with the proposal to make a third trial, that he left the room. The remaining nine, after an earnest debate, resolved, unanimously, to repeat the effort. From that moment the tide turned.

Perhaps some of these courageous nine feared that the third attempt would prove as disastrous as the first and second, but they thought that it ought, nevertheless, to be made; perhaps there were others who expected the success which followed. But could the veil have been lifted from six weeks of the future, how would they have been moved by that which we have witnessed—the swelling emotion, the glad faces, the public rejoicings, which have greeted the victory. They expected, of course, that, when the line was once laid, messages would pass to and fro with instantaneous rapidity; but, however much men may dream of it, the actual occurrence will startle them. Within forty-five days after that meeting of the Directors, news came to London that the Chinese empire, reversing its traditionary policy, and breaking through the prejudices of ages, had made peace with England and France, opening its doors to European intercourse, and, of course, to European culture, but, above all, to the Christian religion. The good news was instantly known in the Western Hemisphere. The imagination is baffled when it tries to picture the journey which the message made. When it left London evening had already come; but it overtook and passed the shadow of the earth, as if that were but a creeping snail, though making daily the circuit of the globe; it darted through the green valleys of England, over Scotch mountains, down beneath the channel to the Irish coast, thence through Ulster and Connaught and Munster to the shores of the Atlantic. Here it dived beneath the ocean, deeper than the valley of Chamouni stands below the summit of Mount Blanc, passing under great ships of commerce and of war, and in an instant arose at the cliffs of Newfoundland; then, quicker than thought, it passed over the morasses and mountains to the Gulf of St. Lawrence, then on through the Gulf, through Nova Scotia, New Brunswick and the Eastern States, to our own doors.

But let us return from this digression to the last expedition. On the 17th of July the squadron departed from Queenstown for the third time. As they passed Cape

Clear, into the Western Ocean, they parted company: but such is the accuracy of modern navigation, that though there was no earthly map or mark to guide them, yet, steering by the compass and the celestial signals, one after the other, all arrived at the appointed rendezvous in mid-ocean.

On the 29th of July the two great ships took their places a short distance from each other. A strong hawser fastened them together. The end of the Cable which the Niagara bore was carried to the Agamemnon and there spliced to the end of hers; it was then lowered into the sea, and the ships moved, each towards its own country, at first creeping slowly till the Cable had sunk far down, and then faster, to a speed of five or six miles an hour.

Let us glance for a moment at the Agamemnon, on her homeward track. She suffered severe weather, and more than once the Cable was in extreme peril. Once, in order to remove a defect in the coil, it was necessary to stop the ship, an operation the most dangerous, for the experience of the two former trials had shown that the insatiable sea will neither give back what it has received, nor allow the supply to cease. But a good Providence watched over the ship, and on the 5th of August she came safely to land.

Let us now return to our own Niagara and her faithful attendant. The Gorgon, herself a ship of 1,100 tons, though but a boat by the side of the Niagara, led the way, because the compasses of the latter were affected by the cable, and the great ship followed close behind. Never was navigator more vigilant and more successful than Captain Dayman. His observations went on by day and by night; as one heavenly body went down and another arose, his instruments were turned to the rising luminary, and he never swerved from the shortest line along the great arc of the circle to the head of Trinity Bay. The Niagara steered by the Gorgon. Her machinery worked with the utmost regularity, never stopping for an instant, and her officers and men were as exact as her machinery. Silence, as far as possible, was enforced, and such light was kept that at night she appeared to the Gorgon to be illuminated. Who can tell what anxious suspense there was in that ship as each hour, each day passed on, increasing the chances of success, strengthening the hopeful, restoring the despondent—what sleepless eyes, what beating hearts were there! As the great ships went forward, from the moment when they disappeared from each other below the horizon, messages were constantly interchanged—ship answered to ship as the hours bore them farther apart and nearer their destination. I scarcely know a dialogue more affecting than that which was held between the Niagara and Agamemnon on this last voyage. At length, on the morning of the 4th of August, under as bright a sky as ever smiled on a great achievement, the headlands of Trinity Bay rose above the sea directly before them.

Then there came out to meet them, and be their pilot into their desired haven, another English ship—the Porcupine—whose Captain, Otter, had so carefully surveyed and so closely watched, that he had not only found all the channels, but had stationed boats to mark the narrowest, and that the ships might be seen far off, had sent sailors into an island of the bay, on which was a high and wooded hill, ordering them to watch day and night, and as soon as the fleet hove in sight to set the wood on fire. The fire was kindled, and the burning hill was at once bonfire and signal for the victorious ships. The bay was so deep that the head of it was not reached till after midnight. There, at five o'clock of the morning of the 5th of August, the end of that mysterious wire was taken ashore; and as soon as it was secured in its appointed station, the brave sailor and humble Christian who commanded the Niagara, in the open air, in the early daylight, while all the gentlemen and seamen bowed their heads reverently, gave thanks to the Almighty for the good voyage ended.

AND THUS WAS THE ATLANTIC CABLE LAID.

The choir then sang the Hallelujah Chorus from Handel's "Messiah."

Hallelujah!
For the Lord God Omnipotent reigneth;
The kingdom of this world is become
The kingdom of our Lord and of his Christ;
And he shall reign, forever and ever,
King of Kings and Lord of Lords!
Hallelujah!

After which the Rev. Dr. Nott, of Union College, offered prayer, as follows:

Almighty God, our Heavenly Father, we thank thee that we have our being in an age signalized by great and repeated triumphs of mind over the material elements of nature. Especially do we thank thee that we are this day permitted to celebrate this mightier triumph in the completion of a medium for the instant interchange of tidings between nations whose intercourse has heretofore been obstructed by the barrier of the mighty ocean. And now, having, in thy good providence, been permitted to pay our willing tribute of respect to those agents whom thou, O Great God, hast employed as instruments in thy hands in bringing about this sublime result, we humbly beseech thee to permit this great assembly, now about to separate, in its own behalf, and in behalf of the dwellers in this and in distant lands, to present, in the name of our common Savior, to our common Father, not a nation's, but a world's acknowledgments, for having inspired the mind of man with the sublime conception, imparted to the intellect of man the science, and to the hands of man the skill, to successfully embed, despite winds and waves, beneath unfathomed waters, that wondrous cord, which, vital with thought, now spans the ocean, over which the free lightnings of Heaven are now constrained to pass, and shall continue to pass, bearing from shore to shore the messages, and doing the bidding of thy creature, man.

And now, our Father, we beseech thee perfect thy begun goodness, and cause these traveled lines of thought to be stretched over continents and islands, and beneath the waters of the sea, till earth's scattered population shall be brought into free and fraternal intercourse, that thus the way be prepared for the diffusion of the knowledge of the Gospel of peace and the introduction of the reign of the Son of God.

While thus supplicating thee to perfect and extend these material cords between the dwellers upon the earth, we thank thee that the medium of intercourse between earth and Heaven, furnished by those chords of redeeming love, extending from thy mercy-seat to the hill of Calvary, where, on behalf of ruined man, thy Son expired, is already perfect, and through which the prayer of faith, as it rises in the heart, and before it has received utterance by the lips, reaches that ear ever open to the supplicant's cry and brings back from thence its sweet response of a reconciled Father's love.

And now, O Thou that hearest prayer, grant, we beseech thee, that the effusions of thy Holy Spirit may accompany the future progressive discoveries of science and triumphs of art, till the dissonance of sin shall have ceased, and harmony be restored throughout thy moral kingdom; That amid the unceasing anthems performed by angels around thy throne may be heard from every hill and valley, over the entire of a redeemed earth, ascending in concordant notes, the Church's Hallelujah:—"Blessing and honor, and dominion, and power, be unto Him that sitteth upon the throne, and to the Lamb, forever and ever." AMEN.

The choir then sang the Doxology:

Praise God from whom all blessings flow,
Praise Him all creatures here below,
Praise Him above, ye heavenly host,
Praise Father, Son, and Holy Ghost.

Upon the conclusion of which, the Rev. David D. Field advanced, and invoked a benediction, and with this the proceedings at the Crystal Palace closed. In half an hour after, the building was completely deserted.

All the presentation addresses having been handsomely engrossed and illuminated on vellum, were attached to sections of the Cable, which were gilded, and had mounted upon one end a solid gold eagle with wings outspread, and upon the other the British lion, also of solid gold.

The following is a description of the gold boxes that were presented: that to Cyrus W. Field, Esq., is four and three-fourths inches long, by two and three-fourths inches wide, and one and a half deep. It weighs fourteen ounces, and is valued at eight hundred dollars. In shape it is a parallelogram, with slightly rounded corners; around the top and bottom, set back a quarter of an inch from the edge, is a single strand of gold cable wound at the corner by twisted yarns, which act as feet, and keep the elaborately engraved surface of the box from contact with surrounding objects.

The ground of the box is brilliantly polished, and all its sides are covered with appropriate designs, engraved with the clearness and delicacy of the finest steel plate. On the lid of the box is a scene representing the splicing of the Cable in mid-ocean, at the rendezvous, the Niagara on the left and the Agamemnon on the right. The ships have just made the splice, and are paying out the slack; the Gorgon and Valorous appear in the background, evidently on the point of starting on their respective piloting missions. The ships are admirably represented, and every attention was paid to detail, so as to render the work historically correct. This design is surmounted with the arms of the city of New York. An inscription is engraved on the polished surface of the lid, on either side of the armorial bearings. The design on the bottom of the box represents the flags of England and the United States arranged in a group above, and twisted in festoons around the sides and below, making a panel within the strand of cable which surrounds the box The panel is divided into two compartments by a coil of cable. In the left compartment are engraved the Federal arms, and in the right one the coat of arms of England. Below, in the centre, is a Morse Telegraph instrument in operation.

On the front side of the box are four emblematic female figures, representing the four quarters of the world. Europe and America have the globe, surmounted by the Cross, in a blaze of glory, between them. Asia and Africa, in humble positions, occupy the corners, but are graciously allowed each to hold one end of a telegraph Cable extended to them by America and Europe.

On the back of the box are seated two female figures, intended to represent Commerce and Science, surrounded by the usual emblematic symbols: Commerce has her ships and cotton bales; Science, her globe and telescope.

On the left end of the box is an engraving of the first meeting, at Mr. Field's house, between the six projectors of the Cable enterprise.

Mr. Peter Cooper sits in the foreground, in front of a table, on which is spread out a chart, one end of which is held by Mr. D. D. Field. Mr. Cyrus W. Field is standing in front of it, and explaining his plan to Messrs. Moses Taylor, Chandler White, and M. O. Roberts, who sit, in close attention, around the table. On the floor is a globe.

On the right end of the box is an engraving of the landing of the Cable at Trinity Bay. The procession of sailors and officers, some twenty-three figures in all, is represented carrying the Cable up a slight ascent. Mr. Field, bareheaded, is in front of the party. Trees and rocks in the background, and the bay, with the ships in the distance, complete the design.

It is altogether a most superb affair, and the conception exceedingly appropriate, while it is the most costly testimonial ever awarded by the municipality of New York.

The three other boxes are similar in shape to that of Mr. Field's, but a little smaller. They measure four inches in length, three inches in width, and one inch in height. They weigh nine ounces, and are valued at six hundred dollars each. The designs are the same except on the ends.

On one end of Captain Hudson's box is a representation of the landing of the Cable at Trinity Bay, the officers and crew of the Niagara standing uncovered, and Captain Hudson offering up prayer. On the other end is a scene illustrative of the taking in of the Cable on board the Niagara.

On one end of Captain Preedy's box is a representation of the landing of the Cable at Valentia Bay (Ireland); and on the other, a scene illustrative of the coiling of the Cable on board of the Agamemnon.

On one end of Mr. Everett's box is a portrait of that ingenious inventor, standing by the paying-out machine, on the deck of the ship, while the Cable is being reeled off. On the other end is a scene in the work-shop, where he is represented studying out the plan of the machinery while it is in the course of construction.

The medals were plain discs of gold, three inches in diameter, each weighing six ounces of solid gold, and valued at three hundred dollars. Each medal is surrounded by coils of Cable, that supply the place of raised edges; the surfaces are highly polished, and the engravings are executed in as high a style of art as those on the boxes.

On the obverse of the medal presented to Captain Dayman, is a fine view of the Niagara going into Trinity Bay, with the Gorgon acting as pilot.

On the obverse of that presented to Captain Aldham is a like view of the Agamemnon, with the Valorous in the background.

The obverse of Engineer Woodhouse's medal is the most elaborately ornamented of the three. It shows upon the lower half a reduced copy of the scene on the lids of the boxes, of the rendezvous in mid-ocean, and the splicing of the Cable. Upon the upper half are two very accurate drawings of the paying-out and brake machinery in use on board the principal ships.

The reverse of all the medals is the same, the designs being the arms of the city of New York, with a space left for an inscription.

The boxes and medals, as well as the mountings of the pieces of Cable around which the addresses were rolled, were the work of the world-renowned jewelers, Messrs. Tiffany & Co., of this city, and were got up in the usual elegant and tasteful style, both as to workmanship and general appearance, for which that firm is so justly famed.

In addition to the medals and boxes furnished for presentation by the city to the chiefs of the expedition, Messrs. Tiffany & Co. were directed to prepare sixty-five other medals, to be presented by the Chamber of Commerce to those engaged in the enterprise. Nine of these medals weighed five ounces each, and were three inches in diameter. These were presented to the parties most prominent in the projecting and carrying out of the undertaking. The happily conceived inscription of award occupied the upper and lower extremes of the reverse face, leaving an intermediate surface, upon which was a finely wrought *bas-relief*, representing the connecting of the Cable in mid-ocean. The design ornamenting the obverse face is essentially emblematic; the centre of this surface bears an exact delineation of the Atlantic hemisphere. Supporting the globe, on either hand, are the figures of Columbia and Britannia. Columbia, garbed as an Indian Queen, bearing upon her right shoulder a well-filled quiver, upon her head an aboriginal diadem of eagle's feathers, in her left hand an olive branch, and in her right an end of the Cable, while at her feet reposes the national bird. Facing the friendly countenance of Columbia stands the British Tutelar, at whose feet the lion peacefully reclines—emblematic of the friendship that has been cemented.

The base upon which this fine sculpture stands is a plain Grecian entablature, beneath which appears the seal of the New York Chamber of Commerce, supported on either side by the shields of the United States and England. A winged figure of Concord hovering over the globe and offering an olive crown to each national genius, completes this fine design.

The smaller medals were each of the weight of three ounces, measuring two and a half inches in diameter, and proportionately thick. The reverse of these is in all respects identical with the large medals, but upon the obverse the artist has introduced the figures of Commerce and Science, as supporting the globe, in the place of Columbia and Britannia. As in the larger medal, the happy idea of surrounding the sphere with a Cable, which is held on each side by the characteristic Genius, is likewise produced in this medal. A Dove, instead of the winged Concord, hovers over the globe, and completes the design. These medals were presented to the other officers on board the vessels engaged in the operation of submerging the Cable.

## THE FIREMEN'S PROCESSION.

The torchlight procession of the New York Fire Department was one of the most magnificent demonstrations this city has ever seen.

While the military were yet marching back to their quarters by companies, after the displays of the afternoon, the fire companies began to march from all parts of the city to the rendezvous on Fifth avenue. Broadway, during the whole period which elapsed from the time the first procession passed, to the hour at which the firemen marched down Broadway, was choked with people and vehicles, so that it was with the greatest difficulty the various fire companies that marched up the thoroughfare made their way. They gradually assembled at the place designated, dressed in fire-cap, red shirt and black pants, according to the printed regulations.

About 7 o'clock the first of the companies arrived, and

all had formed into marching order before the proceedings in the Crystal Palace terminated.

The scene on the avenue was one of great excitement. The sidewalks were crowded, and the companies, as they came on the ground, saluted each other with hearty cheers. The line was formed in eleven divisions, as follows:

First Division, on Fifth avenue, right on Fortieth street.

Second Division, on Fifth avenue, right on Thirty-seventh street.

Third Division, on Fifth avenue, right on Thirty-fourth street.

Fourth Division, on Fifth avenue, right on Thirty-first street.

Fifth Division, on Fifth avenue, right on Twenty-eighth street.

Sixth Division, on Fifth avenue, right on Twenty-fifth street.

Seventh Division, on Fifth avenue, right on Twenty-second street.

Eighth Division, on Fifth avenue, right on Nineteenth street.

Ninth Division, on Fifth avenue, right on Sixteenth street.

Tenth Division, on Fifth avenue, right on Thirteenth street.

Eleventh Division, on Fifth avenue, right on Tenth street.

At ten minutes to 10 o'clock, the assemblage in the Crystal Palace withdrew, and at 10 o'clock the head of the procession started.

Down Forty-second street to Sixth avenue they marched, receiving the city authorities and their guests, who accompanied the procession in carriages. Thence, headed by deputations of policemen from the Tenth and Thirteenth Precincts, commanded by Sergeant Waterbury, they marched down Sixth avenue, then down Broadway; from Broadway, at Union Square, they made a *detour*, and marched round the equestrian statue of Washington, then returned to Broadway, down which they marched to the City Hall park, and passed through the eastern entrance. At one o'clock they were dismissed. The passage of the procession through Broadway was a scene of unrivaled splendor. The sidewalks were crowded with the populace. The balconies, the windows, the house-tops, were filled with people. Almost every house was illuminated, and the decorations that we have already described as being visible during the day, shone with heightened effect and redoubled lustre. The street was clear as noonday, and while the brilliant pageant was passing, afforded a spectacle of popular rejoicing such as has been rarely, if ever, witnessed on this continent. The procession appeared in the following order:

FIRST DIVISION.

Assistant Engineer, John Decker, Marshal.
The Veteran Association of Exempt Firemen.
Ex-Chief and Ex-Assistant Engineers.

Board of Fire Commissioners,

Edward Brown, Wm. A. Freeborn,
William Wright, Andrew Craft,
John W. Schenck:

Officers and Trustees of New York Fire Department Fund.

Dodworth's renowned cornet band, of forty pieces.

Chief Harry Howard, Grand Marshal.

Assistant Engineers John A. Cregier and John Baulch.

Special Aids.

Exempt Engine Company, Zophar Mills, foreman; mustering about seventy exempt firemen, without torches. The machine, which is the largest in the Department, was brilliantly illuminated with a large calcium light.

Marion Engine Company No. 9, James Hayes, foreman; mustering sixty men, marching four abreast, escorting the Fire Department Banner, which was carried by sixteen men, who were alternately relieved as the procession passed along the route of march.

Hook and Ladder Company No. 12, Jas. A. Carolin, foreman; mustering fifty men, headed by Dodworth's Band, of forty pieces, following the Fire Department Banner. The truck had been lately rebuilt, painted, and gilt. On the top of the apparatus was a very brilliant light, furnished by Prof. B. L. Budd, of the Medical College in Thirteenth street. This was relieved by about thirty brilliant truck lights. This is the largest truck in the city.

Hudson Engine Company, No. 1, John Hammell, foreman, turned out sixty men. The engine was decorated with colors. Sixty torches in line.

Marion No. 9, Jas. Hayes, foreman, seventy men ; carrying the magnificent banner of the Fire Department, which was an object of great interest to the immense multitude of spectators.

Friendship Engine No. 12, James A. Carolin, foreman, fifty men. Professor Budd, of the Medical College, in Thirteenth street, furnished the company with a dioptric lens, which was quite a feature of the procession. Signals and lamps were used in abundance, so that Friendship was universally admired.

Hudson Engine No. 1, John Hamill, foreman, sixty men. This machine had a large drummond light, and was handsomely decorated.

Knickerbocker Hose No. 2. This company turned out thirty men, and presented a most creditable appearance,

Independence Hose No. 3, John V. Dalton, foreman, thirty-five men. This cart was appropriately festooned and illuminated.

Whitworth's Band.

Eagle Hose No. 1, Walter Smith, foreman, thirty men. It carried a Riddle's reflector, and looked remarkably well.

Excelsior Engine No. 2, De Lancey W. Knevels, foreman ; sixty men. There was a handsome calcium light on the top of this engine, together with appropriate decorations.

Marion Hose No. 4, Theodore Hiller, foreman, thirty

men. This hose had a splendid plume, and was illuminated with Chinese lamps.

Telegraph Brass Band.

Niagara Engine No. 4, R. W. Adams, foreman, paraded sixty men. Upon the top of the machine was Grant's calcium light, surrounded by eight smaller lights on the arms of the engine.

Protection Engine No. 5, Wm. C. Lyons. This engine was ornamented with various colored ribbons, and the company turned out fifty men.

Mutual Hook and Ladder, No. 1, George A. Hilton, foreman. This truck, which has seen ten years' service, was surmounted by a large gilt eagle. The signal was decorated with ribbons of various colors. Six lamps were suspended between the wheels; and two lighted torches attached to the back of the ladders. On one side was a strip of white canvas, as long as the truck, on which was inscribed in large letters, "The Field is ours." On the other side a similar strip bore the words, "The Atlantic Telegraph, the bridge of thought." From the eagle's beak American flags floated to both ends of the truck.

### SECOND DIVISION.

Assistant Engineer Peter N. Cornwell, Marshal.

Robertson's Band.

Americus Engine No. 6, William Anspake, foreman; eighty men. This machine looked remarkably well. The goddess of Liberty sat in her temple, an eagle soared on the top, flags were suspended from the side, and the engine lit up with variegated lamps.

Croton Hose No. 6, George R. Conner, foreman; thirty men. A large number of fancy lamps were suspended from this cart.

Lexington Engine No. 7, Samuel Cheshire, foreman; sixty men. This engine had a large drummond light, and fifty lamps of various colors.

City Hose No. 8, Charles H. Cornell, foreman; bouquets of real and artificial flowers ornamented this company's carriage. Two twisted coils of blue, white, and red ribbons, emblematical of the cable, were suspended from the rear. The members of this company numbered twenty-six.

Columbian Hose No. 9, John L. Herbell, foreman; thirty men. This hose was illuminated by twenty-five lamps.

Paterson Band.

Ringgold Hose No. 7, A. Winham, foreman; thirty men. This hose cart was beautifully adorned. On the top of the cupola were the names of the Niagara and Agamemnon, while on another part of the cart the names of Field, Hudson, and Everett could be seen.

Water Witch Engine No. 50, paraded forty men. The machine presented a very imposing appearance.

Liberty Hose No. 10, thirty men. Their carriage was very tastefully arrayed with flowers.

Chelsea Hook and Ladder No. 2, Stephen S. Mitchell, foreman; fifty men. This truck was illuminated with lamps of various colors.

Seventy-First Regiment Drum Corps.

New York Hose No. 5, F. Raymond, foreman; twenty-eight men. In front of the carriage were six dark lanterns, and at other parts lanterns.

THIRD DIVISION.

Assistant Engineer Elisha Kingsland, Marshal.

Fort Schuyler Band.

Manhattan Engine No. 8, Robert C. Brown, foreman; seventy men. This machine had an elephant on the top, and splendid signals, and was illuminated by ten large lamps. The men wore drab pantaloons, and presented a fine appearance.

Gulick Hose No. 11, E. G. Robertson, foreman, paraded thirty-six men. Immediately surmounting the hose was a splendid eagle, surrounded with the American colors; in front was a complete model of the Jefferson Market Fire Bell-tower, with a bell in the centre, and the whole tower illuminated.

East River Engine No. 17, Christopher Reynolds, foreman; sixty men. This machine was handsomely ornamented with lanterns and flags, and a portion of the Atlantic Cable, furnished by the Chief Engineer of the Niagara, was placed in a conspicuous position.

Jackson Hose No. 13, Archibald Irvine, foreman ; thirty men. This cart carried a portion of the Atlantic Cable, and was beautifully decorated with lamps of various colors.

Excelsior Hose No. 14, thirty men. They presented a very imposing appearance.

Eleventh Regiment Band.

Lafayette Engine No. 19, fifty men. The machine was richly decorated with some sixty odd lights.

Tompkins Hose No. 16, Alex. Ferris, foreman ; thirty men. This cart was illuminated with lamps.

Union Eng'ne No. 18, Jas. Conolly, foreman; sixty men. This engine was illuminated with Roman candles, and looked well.

Phenix Hook and Ladder No. 3, James Galway, foreman; paraded fifty men. A drummond light adorned the centre of the truck, accompanied with forty lights.

Clinton Hose, No. 17, Laurence Dalton, foreman ; thirty men. This hose was tastefully festooned.

### FOURTH DIVISION.

Assistant Engineer W. T. Mawbey, Marshal.

Shelton's Band.

Oceanus Engine No. 11, James Wildley, foreman ; paraded fifty men. A large drummond light surmounted the box, in front of which was a deer's head.

Franklin Hose No. 18, D. J. Conley, foreman. A large gilt eagle surmounted this carriage, and held in its beak a wreath of flowers; an American flag was wrapped around the hose reel. The carriage was illuminated by thirty colored lanterns. In front was a transparency bearing the following:

1752 FRANKLIN. 1752

"Franklin reared the kite that found the spark;
Morse sent it on from zone to zone;
Field spanned old Ocean's briny depths:
America, the work's thine own."

This company turned out forty men.

America Hose No. 19, Walter W. Adams, foreman; paraded thirty men. Over the jacket was the following transparency:

"Electricity—Franklin bottled it and left it a legacy to this country. Morse uncorked it and invited all the nations to partake thereof."

Upon the reverse side was:

"AMERICA HOSE NO. 19."

Turl's Brass Band.

Harry Howard Hook and Ladder No. 11, Charles N. Kent, foreman; fifty men. This truck was an object of attraction, and called forth the plaudits of the spectators, and deservedly so.

Phenix Hose No. 22, paraded thirty odd men; they were dressed in the regular uniform, and, with their carriage, attracted considerable attention.

Protector Engine No. 22, Caleb Sears, foreman; an illuminated signal in the shape of a fire hat surmounted this engine. A number of lamps were strung along the brakes, which produced a very pleasing effect. No. 22 paraded fifty members.

Thomas Manahan's Brass Band.

Eagle Engine No. 13, John Healy, foreman. American flags were wrapped around the reels of this engine in front and rear. A number of lamps were hung on the brakes or handles. The company mustered sixty men.

Humane Hose No. 20, Jared A. Timpson, foreman.—This company mustered thirty men. The front of the carriage was decorated with American and British flags intertwined.

Eagle Hook and Ladder No. 4 paraded thirty-five men, and had their truck very finely decorated.

Band.

Washington Engine No. 20 John Roberts, foreman.—This engine held between its "arms" a model of the Niagara, about three feet long, the work of one of the members. On the front of it was a drummond light. Fifty members turned out.

Hudson Hose No. 21 paraded thirty men; they attracted considerable attention along the line on Broadway.

United States Engine No. 23 was also out with sixty men. This engine was illuminated with lanterns.

## FIFTH DIVISION.

Assistant Engineer Timothy L. West, Marshal.

Stewart's Band.

Columbian Engine No. 14, R. Rogers, foreman. This engine was decorated with artificial flowers. From its front the figure of an elephant was suspended, and from the back hung the representation of a lion. Eight members, bearing lights, marched by the side of the machine. The company turned out sixty men.

Perry Hose No. 23, A. V. Davison, foreman, paraded thirty men. In the centre of the carriage was a mammoth cock, with the words "victory" flying from its mouth; a large feather adorned the front of the carriage.

National Hose No. 24, S. Burhans, Jr., foreman, paraded thirty men; decorated with lights and lanterns.

Union Hook and Ladder No. 5, Adam Hipp, foreman, fifty men. This ladder company looked well, but had no extra adornment.

Jackson Engine No. 24; paraded about sixty men. Their engine attracted much attention.

Edward Manahan's Brass Band.

Fulton Engine No. 21, James M'Cullough, foreman. In front of this engine, which was illuminated by ten camphene lamps hung around it, and by a number of torches, was a large American flag. Fulton engine turned out the full compliment of sixty men.

United States Hose No. 25, Hugh Gallagher, foreman. The American and British flags waved over this hose carriage. The British flag used on this occasion was presented to the company by the garrison of Windsor, Canada. A large and fierce-looking wild cat, captured expressly for and presented to the company by their friends in Detroit (Michigan), was perched on the hose reel, and held a lantern in his mouth. This company turned out thirty men.

Williamsburgh Brass Band.

Cataract Engine No. 25, William Lamb, foreman, sixty men. This engine was also beautifully illuminated, and surmounting it was a buck's head enveloped in flags.

Fourth Regiment Brass Band.

Howard Engine No. 34, J. L. Coe, foreman, paraded sixty men. The engine was very tastefully decorated with flags and lights—twenty-five lanterns being hung at various portions of the machine.

Rutgers Hose No. 26, James M. Petty, foreman; thirty men. A large number of lamps of various colors were suspended from this hose in a tasteful manner.

Jefferson Engine No. 26, John Ford, foreman; fifty men. This Engine is herself an artistic and beautiful piece of mechanism, and required no decorations. "Beauty, when unadorned 's adorned the most."

## SIXTH DIVISION.

Assistant Engineer James F. Wenman, Marshal.

Governor's Guard Band.

Amity Hose No. 38, Isaac M. Barnby, foreman, paraded twenty-six men. The carriage was adorned with a large map of the world, over which was a massive colored light representing the globe, with the word "Amity" at the top. Each man also carried a lantern in his hand.

Neptune Hose No. 27, John H. Corballis, foreman. The hose carriage belonging to this company was surmounted by a silver eagle; it was plainly, but tastefully, decorated. The company was represented in the procession by thirty members.

Metamora Hose No. 29, J. E. Conklin, foreman, paraded thirty men. Carriage well trimmed with flowers and ribbons.

Hook and Ladder No. 6, James Kellock, foreman, fifty men. In the centre of the truck was a large field piece, in front of which was the coat of arms of the Fire Department.

Guardian Engine No. 29, Eli Bates, foreman, paraded sixty men; engine tastefully decorated, accompanied with lights.

Knickerbocker Band.

Knickerbocker Engine No. 12, Jacob W. Cooper, foreman; fifty men. A live eagle, three and a half feet in height, was placed on the top of this engine and attracted much attention.

Laurel Hose No. 30, James H. Arnold, foreman; thirty men. This cart was lit up by red and white colored lanterns.

Index Hose No. 32, William Holden, assistant foreman in command; thirty men. It was decorated with flags and beautiful flowers.

SEVENTH DIVISION.

Assistant Engineer Edward W. Jacobs, Marshal.

Connell's Brass Band.

Black Joke Engine No. 33, was beautifully adorned with flags and lit up with variegated lamps. Sixty men appeared and contributed no little to the display.

Warren Hose No. 33, John D. Craft, foreman; thirty-five men. The carriage was very tastefully decorated with flags.

Lafayette Hose No. 34, John Irvine, foreman; thirty men: This cart had a large plume on top, and was enveloped in flags.

Chatham Engine No. 15, David S. Baker, foreman, seventy men. The engine was not adorned, but signals were suspended from it, and the company made an excellent turn-out.

Columbus Engine No. 35, paraded fifty men, and made

a very fine appearance; the machine was trimmed off with a great deal of taste.

Baltic Hose No. 35, James H. Bell, foreman, paraded thirty-six men. Carriage trimmed with American flags.

Empire Band.

Empire Hook and Ladder No. 8, John C. Everett, foreman; fifty men. This ladder had a large number of lights and forty-eight torches.

Equitable Engine No. 36, was festooned with flowers, adorned with flags, and brilliantly illuminated. They turned out sixty men.

Empire Hose No. 40, William Evans, foreman, paraded thirty men. Upon the top of the reel jacket was a large drummond light. On either side of the carriage was the likeness of J. A. Cregier, assistant engineer, James L. Miller, David Mulligan and John Kittleman. No. 40's carriage was about one of the finest in the procession; being new it attracted no little attention.

## EIGHTH DIVISION.

Assistant Engineer G. Joseph Ruch, Marshal.

Wallace's Band.

Southwark Engine No. 38, George T. Alker, foreman, made a splendid display. The engine, which is a first-class one, was drawn by four of Adams' Express Company's horses—known as the "string team." The horses' heads were decorated with red, white and blue plumes. The company turned out the full number allowed, and were accompanied by several honorary members of many years' standing. They were lavish of Roman candles and Bengal lights, and were headed by Wallace's band.

Naiad Hose No. 53, Wm. H. Shumway, foreman. In front of the hose carriage was a transparency, representing the two Atlantic telegraph stations and the telegraph itself on its couch beneath the waves. Six naiads were pictured giving expression to their joy at the presence among them of the illustrious stranger. On the rear was another transparency with the motto, "The Electric Spark —one we would not extinguish." Sixteen flags of different nations, among which were the American, English, French and Belgian standards, were festooned at each corner. In pans attached to the centre of the panel or reel, Bengola lights were kept continually burning. The company turned out thirty members.

Tradesmen's Engine No. 37. This machine was magnificently festooned and brilliantly illuminated. Sixty men marched in the parade.

Madison Hose No. 37, Simon V. Wooley, foreman; thirty men. This cart was adorned with flags and lit up with various colored lamps.

Lady Washington Engine No. 40, Joseph H. Hutton, foreman. On the top of this engine was a large carved eagle holding in its beak the American and English flags. It was illuminated by several lamps and a drummond light. Sixty members had turned out.

Band.

Washington Hook and Ladder No. 9, John H. Forman; fifty men. This truck had an immense drummond light, a large flag and countless colored lamps.

Pacific Engine No. 28, Samuel M. Simpson, foreman; fifty men. This machine attracted the attention of the crowd; an immense calcium light, and one hundred and fifty lamps, illuminating it, together with banners.

Metropolitan Hose No. 39, Hugh Hanley, foreman; thirty men. This hose was illuminated with lamps.

Manhattan Engine No. 43 also attracted considerable attention, the machine being appropriately adorned and lit with a great number of colored lamps. Fifty men paraded, in the regular uniform, and looked well.

Pioneer Hose No. 43; was also tastefully decorated. Thirty men paraded.

## NINTH DIVISION.

Assistant Engineer John Brice, Marshal.

Fifty-fifth Regiment Band.

Empire Engine No. 42, Richard P. Moore, foreman. This engine was lighted with twenty Chinese lanterns of various colors. The members of the company paraded in a new uniform, viz.: gray shirts, black pants and neckerchiefs; they numbered fifty men.

Mazeppa Engine No. 48, Jeremiah Foley, foreman; sixty men. This engine was decorated with flags and illuminated with lamps.

Mazeppa Hose No. 42, John Lee, foreman ; thirty men. This cart carried a large drummond light and a quantity of lamps.

Narragansett Hook and Ladder No. 10. The truck looked well, and the members of the company bore themselves handsomely.

Franklin Engine No. 39, looked exceedingly well, and her members mustered in respectable numbers.

Adkins' Washington Brass Band.

Clinton Engine No. 41, Anson Alaire, foreman ; sixty men. This engine was polished in a beautiful manner, and attracted considerable attraction.

Alert Hose No. 41, Wm. McLaughlan, foreman ; thirty men. This hose cart was decorated with American flags, and illuminated with lamps.

Washington Irving Hose No. 44. The members of this company presented an imposing appearance ; their carriage was tastefully decorated.

Excelsior Band.

Valley Forge Hose No. 46, Edward L. Cobb, foreman ; thirty men. This hose was brilliantly illuminated, and looked beautiful.

Mechanics' Hose No. 47, Cornelius N. Wright, foreman ; thirty men. This hose cart was not decorated in an unusual manner.

Mazeppa Engine No. 48, Jeremiah Foley, foreman. This company turned out sixty men, and their machine was decorated with flags and illuminated by lamps of various colors.

TENTH DIVISION.

Assistant Engineer Daniel Donovan, Marshal.

Bronck's Band.

Marion Hook and Ladder No. 13, E. A. Gregory, foreman ; fifty men. This company had a large lamp, an eagle and variegated lamps.

Americus Hose No. 48 had thirty men, and looked admirably.

Pocahontas Engine No. 49. This engine was splendidly decorated.

Relief Hose No. 51 paraded with thirty men ; the cart was beautifully illuminated.

Sixty-ninth Regiment Drum Corps.

Hope Hose No. 51. This cart was very appropriately illuminated and decorated.

Yorkville Band.

Aurora Engine No. 45. This machine was also very handsomely decorated with flags and signals, and was brilliantly illuminated. Sixty men paraded.

Eureka Hose No. 54. This company's carriage was ornamented with a number of bouquets of natural flowers, several wreaths of artificial flowers, and surmounted by two American flags. The members in the cortege numbered forty men.

Lone Star Engine No. 50, sixty men, was brilliantly illuminated with a profusion of variegated lamps, the company having provided an extra number over the quantity appropriated by the Corporation.

Undine Hose No. 52. This machine was handsomely fitted out with flags and lamps, and the men marched well.

Columbian Hook and Ladder No. 14, fifty men. This truck was decorated with festoons, and illuminated with a large number of lamps.

### ELEVENTH DIVISION.

Assistant Engineer Wm. Hackett, Marshall.

Dodworth's Second Band.

Baxter Hook and Ladder No. 15. This truck was draped with American flags, and surmounted by a very large gilt eagle. It was illuminated by several blue, white and red lanterns, and a drummond light blazed forth on the fore part of the truck. The company numbered fifty men, and was preceded by Dodworth's Second Band, of thirty instruments.

Harry Howard Hose No. 55 paraded thirty men. Upon the carriage were ten beautiful lights, each member also carrying a lantern in his hand.

Mutual Engine No. 51; sixty men. A splendid wreath of flowers was placed in front of this engine, and in the centre hung a huge lamp, while the American eagle was placed in the rear and from its mouth six feet of the Atlantic Cable was quite visible, affording Young America an opportunity to let off a stock of adjectives. Samuel G. Jackson, a member of the company, was on board the Niagara and furnished the Cable.

Nassau Hose No. 56. This company was out in good force, and both the members and their carriage looked well.

Paulding Hose No. 57, John J. Reed, foreman, paraded thirty men. Their carriage was beautifully decorated with flowers and lanterns.

New Jersey Brass Band.

M. T. Brennan Hose No. 60, Walter Roche, foreman. In front of this engine were American flags intertwined, and on the top there was a powerful drummond light. Thirty members of this company paraded.

Merchant Hose No. 58. The members of this company made a very creditable turn-out, and were loudly applauded by their friends.

This closed the procession, and, with it, closed one of the most imposing displays of this age and country.

### THE FIREWORKS.

All the day, from the moment the sun marked early breakfast-time until he set in a cloudless West, workmen were putting up curious sticks and odd cross-bars along the front of the dilapidated City Hall. These sticks and bars were the foundations of a cable and two ships, with all the accompaniments necessary to give the telegraph a fiery salute. At nightfall the work was finished, or rather so far finished that it was ready to be begun when eight o'clock came.

While the firemen were getting their ropes in readiness, and when their lamps were trimmed and burning, Mr. Lilienthal's men applied the torch to combustibles in front of the City Hall, and a conflagration was the result. First came a " Revolving Globe." This had double revolutions, spun round a spindle, and rotated in company with a wheel that spun about in a contrary way; altogether good, though two brilliant things were at cross-purposes. Colored fires and lance-work formed the features of this initial piece. And, when one was done at the west wing of the Hall, another just like it fizzed away at the east wing. So nobody was favored above his neighbor, and all saw and shouted an approval.

For a few minutes after the " Revolving Globe" came to a black finish, that immense, prodigious mass of people stood patiently, while a little shower of rockets went up towards Heaven. Then came piece No. 2—a *Polka Quadrille.* This piece was eight feet long—its main part a large wheel, bearing upon the ends of the spokes sundry small Saxon wheels revolving rapidly in opposite directions, with red fires changing to green, and a general effect altogether very excellent.

There was another pause, and then *Saturn and his Satel-*

*lites.* The centre of this piece was stationary. Small Saxon wheels revolved, displaying colored fires, and the whole ended with colored fires and brisk cannon discharges.

The fourth was a *Compound Saxon Wheel*—diameter five feet, the spokes containing small Saxons, and ending, not by cannon discharges, but with the evolutions of big wheels and little wheels.

Four grand pieces had now been discharged; all good. The *finale* was yet to come. Before, between all the pieces, and after everything but this final "blaze of glory," the men who were perched upon the very top of two high platforms, placed one at each end of the Hall, threw high into the air an almost endless shower of the finest rockets, bombs and projectiles of all sorts, that burst a mile or so above, and sent down stars and golden rain to celebrate the last of their existence. Mr. Lillienthal certainly did great things in the rocket line alone. The profusion was something extraordinary. The air was full of streaming lines of light, trails of fiery serpents and clusters of stars. The scene, with these buzzing and flashing, and the dense mass below and all around shouting, and a path of silvery light passing over the park trees from a calcium light over in Broadway, was decidedly impressive.

The emblematic piece, which ended the show, was magnificent. It was an allegorical representation, founded on the Cable—giving thanks for the Cable—illustrating and illuminating the Cable. The piece occupied the entire front of the City Hall—wings and all. Its total length was one hundred and eighty feet. At either end was a full rigged ship, one the Niagara, the other the Agamemnon.

Each vessel was twenty feet from stem to stern; and the height the same up to the peak of the mainmast. An entablature, one hundred and twenty feet long, stretched along the front of the Hall, from wing to wing. This entablature was supported by six pillars, twenty feet in height. Three of these pillars were on either side of a rotating Maltese cross. Along the face of the structure ran, in letters of fire, this inscription:

"THE LAST NAVAL ENGAGEMENT BETWEEN COLUMBIA AND BRITANNIA—A TIE."

Over this was old Father Neptune, reclining in his shell in mid-ocean, his trident over his shoulder, and one hand uplifted, holding a lighted cord—the Cable. The Cable ran east to "our Eagle," west to "their Lion"—a surly fellow, with his paw, holding fast his end of the cord. Behind the eagle stood a herald, trumpeting fame; behind the lion stood another herald, trumpeting more fame. Over all this show sprung an arch, bearing, in letters of gold, these words:

"GOD HAS BEEN WITH US: TO HIM BE ALL THE GLORY."

The Cable springing from mid-ocean, the ships that payed out the Cable, the pious telegram sent over the wires to announce the great result, the double-barreled pun in the first inscription, the grand demonstration that wound up the whole—elicited continuous shouts of approbation. The entire populace—and how many tens of thousands there were packed into the park, we don't like to venture to peril our veracity by stating—gave a universal verdict on this occasion, that the Cable was a great event.

Such an impenetrable mass of human beings it could hardly be said had ever assembled before within the city of New York. The crowd was dense enough during the day, but night brought people to the City Hall by tens of thousands. They were crammed together under the trees, straining every muscle to obtain a glimpse of the display; and far away, toward the Astor House, the throng extended in undiminished density. It is certainly safe to say that at least one hundred thousand people were present. Over this huge throng, which swayed to and fro with every impulse, the blue and red lights ever and anon shed an unearthly glare. It was a deeply impressive sight; and the "sea of upturned faces," reddened by the glow, were grander far to witness than any pyrotechnic display that could be invented. At length the exhibition was brought to a close, with the piece which we have already described. The people got into motion, and as they slowly moved from the park, in the most perfect order, a parti-colored illumination once more lightened up every place—and then all was dark once more.

And thus ended the first day's celebration of the laying of the Atlantic Cable.

## THE MUNICIPAL BANQUET.

In accordance with the arrangements made by the Joint Committee of the Common Council, the grand municipal banquet given by the Corporation to Cyrus W. Field, Esq., and officers of Her Britannic Majesty's steamship Gorgon and United States steam frigate Niagara, in commemoration of the laying of the Atlantic Cable, came off at the Metropolitan Hotel, Broadway, on the evening of September 2d. The dining-hall was elaborately and tastefully decorated with flags and devices appropriate to the occasion; on the left side of the head of the table of honor was hung a copy of Winterhalter's celebrated painting of Queen Victoria, and on the other side a painting of President Buchanan. Between the two was a portrait of Mr. Cyrus W. Field. Underneath the latter was the inscription:

"PEACE ON EARTH, GOOD-WILL TO MEN."

Over the portrait the flags of the United States and Great Britain were gracefully interfolded, and a portion of the Submarine Cable was looped to the ceiling above it and carried in festoons throughout the room. From the side walls hung out the flags of various nations, and from the central chandelier were spread over the ceiling the flags of England, France, Russia and the United States. In the spaces between the windows were hung strips of canvas bearing respectively the names of Franklin, Bright, Berryman, Everett, Woodhouse, Otter and Dayman. In the centre of the room, opposite the orchestra, was a portrait of Morse, with the inscription:

"THE ELEVATION OF MANKIND."

"THE ELECTRIC TRIDENT, WAKING TO LIFE THE UNCIVILIZED WORLD."

On the opposite side of the room, in front of the music gallery, was the inscription in the centre:

"THE ATLANTIC CABLE.

AN AUDIENCE OF THE WORLD, WITH HOSANNAS, WELCOMES GENIUS."

And at either side the inscriptions:

"THE OCEAN ECHO.

"SILENT IN ITS COURSE BENEATH THE WAVES, YET ELOQUENT IN ITS THUNDERS THROUGH TWO HEMISPHERES."

"THE ELECTRIC SPARK.

FRANKLIN PLAYED WITH THE CLOUDS—AMERICAN GENIUS GAVE SPEECH TO THE OCEAN."

The following is a copy of the bill of fare. It was headed by an engraving of telegraphic operators sending and receiving messages on either side of the Atlantic, and its border was a representation of the Atlantic Cable.

OYSTERS ON THE HALF-SHELL.

*Soups.*

Green Turtle. Gumbo, with rice.

*Fish.*

Boiled Fresh Salmon, lobster sauce.
Broiled Spanish Mackerel, Steward's sauce.

*Boiled.*

Turkey, oyster sauce. Leg of Mutton, caper sauce

*Roast.*

Ribs of Beef. Young Turkey.
Lamb, mint sauce.
Ham, champagne sauce. Chickens, English sauce.

*Cold Dishes.*

Boned Turkey, with jelly.
Pâtes of Game, with truffles, Chicken Salad, lobster sauce.
Ham, sur socie, with jelly.

*Entrees.*

Tenderloin of Beef, larded, with mushroom sauce.
Lamb Chops, with green peas.
Chartreuse of Partridges, Madeira sauce.
Forms of Rice, with small vegetables.
Timbale of Macaroni, Milanaise style.
Wild Ducks, with olives.
Breast of Chickens, truffle sauce.
Soft Shell Crabs, fried plain.
Stewed Terrapin, American style.
Squabs, braisees, gardener's sauce.
Sweetbreads, larded, with string beans.
Fricandeau of Veal, larded, with small carrots.
Flounders, stuffed, with fine herbs.
Reed Birds, Steward's sauce.
Broiled Turtle Steaks, tomato sauce.
Croquettes of Chickens, with fried parsley.
Tenderloin of Lamb, larded, poivrade sauce.
Pluvier, on toast, Italian sauce.

*Relishes.*

Pickled Oysters. Raw Tomatoes.
Spanish Olives. Celery. Currant Jelly.

*Game.*

Partridges, bread sauce. Broiled English Snipe.

*Vegetables.*

Boiled and Mashed Potatoes.
Sweet Potatoes.
Stewed Tomatoes. Lima Beans.

*Pastry.*

Apple Pies. Peach Pies. Custard Pies.
Pineapple Pies. Pumpkin Pies.
Plum Pies.

Plum Pudding.
Fancy Ornamented Charlotte Russe.
Maraschino Jelly.
Fancy Fruit Jelly.
Pineapple Salad.
Gateaux Neapolitan style.
Peach Meringues.
Cabinet Pudding.
Madeira Jelly.
Punch Jelly.
Fancy Blanc Mange.
Spanish Cream.
Swiss Meringues.
Champagne Jelly.

*Confectionery.*

Meringues, a la crême, vanilla flavor.
Rose Almonds.
Fancy Lady's Cake.
Ornamented Macaroons.
Mint Cream Candy.
Butterflies of Vienna Cake.
Savoy Biscuit.
Variety Glace Fruit.
Quince Soufflée.
Vanilla Sugar Almonds.
Fancy Diamond Kisses.
Preserved Almond Kisses.
Dominos of Biscuit.
Fancy Variety Candy.
Roast Almonds.
Conserve Kisses.
Chocolate Biscuit.

VANILLA ICE CREAM.

*Dessert.*

Almonds, Peaches, Pecan Nuts. Grenoble Nuts.
Hot House Grapes. Citron Melons.
Raisins. Bartlet Pears. Filberts.

COFFEE.

Among the confectionery ornaments of the table were pieces representative of

Queen Victoria, of Great Britain.

James Buchanan, President of the United States.

Cyrus W. Field, with his Cable.

Professor Morse, as Inventor of the Telegraph.

Dr. Benjamin Franklin.

The operative Telegraph of the Metropolitan Hotel.
The Niagara Man-of-War of the United States.
The Agamemnon and Niagara paying out the Cable.
Cyrus W. Field, surrounded by the flags of all nations.
The Coat of Arms of all nations, on a pyramid.
Pocahontas, with real American design.
Grand ornamented Fruit Vase.
Sugar Tower, with variety decorations.
Fruit Basket, supported by dolphins.
Fancy decorated Flower Vase.
Lyre, surmounded with cornucopia of Flowers.
Pyramid of Cracking Bonbons.
Sugar Harp, with floral decorations.
Scotch Warrior, mounted.

| | |
|---|---|
| White Sugar Ornament. | Ornamented Sugar Tower. |
| Temple of Liberty. | Temple of Music. |
| Frosting Tower. | Flower Pyramid. |
| Tribute Temple. | Pagodi Pyramid. |
| Ethiopian Tower. | Floral Vase, decorated. |
| Frosting Pyramid. | Mounted Church. |
| Chinese Pavilion. | Triumphant Temple. |
| Variety Pyramid. | Fancy Sugar Temple. |

Temple of Art.

The Chair was occupied by his Honor Daniel F. Tiemann, Mayor of the city of New York. At his right sat Cyrus W. Field, Captain Dayman, of the Gorgon; Mr. Everett, United States Navy; Mr. E. W. Archibald, British Consul; Governor King, Archbishop Hughes, Rev. Dr. Field, father of Cyrus W. Field; Rev. Dr. Adams and Wilson G.

Hunt. At his left sat Lord Napier, British Minister to Washington; Captain Hudson, of the Niagara; Rev. Dr. Ogilby, of Trinity Church; Peter Cooper, the Bolivian Consul, General Sandford and Lieutenant M. F. Maury.

Besides the distinguished and prominent citizens of New York who were present, there were also present as guests, Count Viscomte, B. Mitchell, T. N. Roberts, and F. B. Butler, of H. B. M. steamship Gorgon; Captain W. Rose Hall, and Lieutenants Hugh Davis, L. A. Bell, Josiah C. Cole, A. H. Trainer, P. C. Johnstone, and A. T. Kingstone, of H. B. M. ship Indus; Lords Ribblesdale, R. Grosvenor, and F. Cavendish, Hon. John M. Young, of Canada, Henry T. O'Reilly, Señor Ferrera, H. C. M. Minister, Hon. J. E. Ward, of Georgia, Hon. Henry R. Jackson, United States Minister to Austria, Isaac N. Marks, of New Orleans, ex-Governor Rodman M. Price, of New Jersey, Señor Aguilar, the Brazilian Consul, Robert Bunce, British Consul for North and South Carolina, Mr. Edwards, British Vice-Consul, Professor Bache, of the Coast Survey, Captains Shestakoff, Schwartz, Ivanschemko, Salmenioff, and Belaventz, of the Russian Navy, and others of distinction.

The Presidents, members, and Secretaries of the Boards of Aldermen, Councilmen, and Supervisors of the city, members of the Judiciary and of the bar, also participated in the celebration.

Many of the most illustrious and eminent among the invited guests, were unavoidably precluded from being present, much to the regret of the Committee.

Some of the letters of regret, most aptly expressing the feelings animating the community, and the unanimity of appreciation on the part of the representatives of other nations upon this occasion, are herewith submitted.

[From His Excellency JAMES BUCHANAN, President of the United States.]

WASHINGTON, August 28, 1858.

MY DEAR SIR—I have been honored by the very kind invitation of the Mayor and Common Council of the city of New York to be present at the festivities to take place on the 1st proximo, and at the municipal dinner to be given on the 2d, in celebration of the successful laying of the Atlantic Telegraph Cable.

I cordially reciprocate their congratulations upon the success of this wonderful enterprise. It is the miracle of this age of miracles. From the very beginning, my sympathies were warmly enlisted in its favor, and my faith was strong on its eventual success. Mr. Field inspired me with a portion of his own ardent spirit in the cause. Science taught us that it was possible; and this being the case, I knew that what was possible, would be, at last, accomplished by the united skill, energy and perseverance of Englishmen and Americans. May it prove a bond of perpetual friendship between the kindred nations!

No man can anticipate what will be its effects upon the people of the two nations and upon the world. Like all great and novel enterprises it may, in the commencement, have some drawbacks. I am firm, however, in the faith that so mighty an agent for the extension of commerce between the nations of the earth, and for promoting the union of the world in one vast brotherhood is designed by an over-ruling Providence to confer lasting blessings and benefits on all mankind. Let us then rejoice!

Such being my sentiments, I deeply regret that the pressure of public business will deprive me of the privilege of uniting with my fellow-citizens of New York in commemorating this great event.

Yours, very respectfully,
JAMES BUCHANAN.

THOMAS McSPEDON, Esq., *Chairman, &c., &c.*

---

[From ex-President MARTIN VAN BUREN.]

LINDENWALD, August 30, 1858.

DEAR SIR—I have had the satisfaction to receive, through you, the congratulations of the Mayor and Common Council of the City of New York upon the successful completion of the attempt to lay the Atlantic Telegraph Cable. I beg you to assure them that I reciprocate, with all my heart, the feelings they have expressed.

The great transaction of which they speak is one of those remarkable events which address themselves too forcibly to the hearts and minds of all, who feel and think, to leave it in the power of words to increase their respect or thankfulness for its accomplishment. It speaks its own eulogy.

The Common Council have, also, through the same channel, done me the honor to invite me to attend the festivities that are to take place, by their authority, in commemoration of the event; including a municipal dinner to be given to Mr. Field, and to the officers of H. B. M. Steamship Gorgon and the U. S. Steamer Niagara.

Concurring very fully with the Common Council in their appreciation of the meritorious services of the recipients of their favorable notice, it would give me much pleasure to take part in a festival designed to do them honor; but I regret to say that I am constrained to deny myself that gratification. Those who have contributed to the accomplishment of this great object, whether through scientific improvements, or by their persevering efforts in the successful application of them, are justly entitled to the respect and gratitude of mankind; and the Common Council do themselves much honor in bestowing testimonials of theirs, as far as they have found it practicable, upon those who have assisted in the advancement of this great work.

Please, sir, to present to His Honor the Mayor, and to the Honorable Council my respectful acknowledgments for this proof of their regard, and believe me,

Very respectfully, your obedient servant,

M. VAN BUREN.

THOMAS MCSPEDON, Esq.,

*Chairman, &c.*

---

[From the Hon. LEWIS CASS, Secretary of State.]

WASHINGTON, August 28, 1858.

SIR—You will oblige me by tendering to the Joint

Committee of the Common Council my thanks for the honor they have done me, by their invitation to attend the festivities about to take place, in the city of New York, in commemoration of the successful laying of the Atlantic Telegraph; an event which commands, as it merits, the universal admiration of the American people. I cannot be with you upon that interesting occasion, for I shall be detained here by my public duties; but, though absent, I shall fully participate in all those sentiments of national pride and of hope which the fortunate accomplishment of this great enterprise is so well calculated to excite.

With great respect,

I am, sir,

Your obedient servant,

LEWIS CASS.

THOMAS McSPEDON, Esq.,

*Chairman, &c.*

---

[From the Hon. HOWELL COBB, Secretary of the Treasury.]

WASHINGTON CITY, August 28, 1858.

DEAR SIR—I acknowledge and cordially reciprocate the congratulations of the Mayor and Common Council of the city of New York "upon the successful laying of the Atlantic Telegraph Cable," which, by their direction, you have communicated to me.

I regret to say that my official engagements will prevent the acceptance of their invitation to be present at the proposed "festivities," on the first of September next.

It is difficult to realize the fact, that this extraordinary enterprise has been crowned with success. Reflections upon the progress of science and the energy of man fail to satisfy and quiet the mind startled by the contemplation of this wonderful result.

In the wisdom and power of God alone we read a solution of what, in any other view, would be incomprehensible.

Let us, then, do honor to the men whose science and energy, under the guidance of Providence, have thus marked the time in which we live as an era in the history of the world's progress.

All speculation is at fault in anticipating the ultimate results which may flow from the success of this enterprise. Time alone can develop them.

I am, very respectfully,

Your obedient servant,

HOWELL COBB,

*Secretary of the Treasury.*

THOMAS McSPEDON, Esq.,

*Chairman, &c., New York City.*

---

[From the Hon. J. THOMPSON, Secretary of the Interior.]

DEPARTMENT OF THE INTERIOR,

WASHINGTON, Aug. 30, 1858.

SIR—Nothing but the imperative demands of public business would prevent my acceptance of your kind invitation to join the Mayor and Common Council in the festivites to take place "in commemoration of the successful laying of the Atlantic Telegraph Cable," on the 1st prox.; as also at the municipal dinner to be given at the Metropolitan Hotel on the evening of the 2d.

I should do myself injustice to deny my earnest desire to be with you, to express my gratification at an event so important, so wonderful, and so unexpected, even by the most hopeful and intelligent. But the work, which is to mark and distinguish the age in which we live, is accomplished, and those who performed it richly merit the consideration and distinction which the Mayor and Common Council of your city so cheerfully and so justly accord them. The country feels, and will readily acknowledge, the honor which has been gained for it. The civilized world will be electrified, and, leaving the dead past to bury its dead, will awaken to new enterprises of "vast pith and moment," by which the triumphs of peace and science, now in obedience to that first great command of God to man, to occupy and subdue the earth, will render contemptible the achievements of arms.

The moral sense of the world, rendered more delicate by the rapid circulation, along this great artery, of the religious sentiment of Christendom, will be shocked at the ruffian nation which shall rely upon powder and lead to enforce respect and deference for its rights. On the 5th of August, 1858, an era was established which divides the past from the future. The order was then given to all civilized nations to ground arms against Christian man, and that, henceforth, the engines of war for man's destruction should be employed only against the heathen and the barbarian.

The completion of the magnificent fortress at Cherbourg is an event which belongs to a barbaric age. Henceforth, the kingdoms of the earth must learn righteousness and do justice. Henceforth, the wrongs which the weakest people shall suffer will touch a chord, which, with the lightning's flash, will vibrate through the world. Let us, then, rejoice and make glad, for a truly great and wonderful work has been accomplished in this our day.

I regret that I cannot be personally with you.

With high regard,

Your obedient servant,

J. THOMPSON.

To THOMAS MCSPEDON,

*Chairman, &c.*

---

[From the Hon. A. V. BROWN, Postmaster-General.]

WASHINGTON, September 1, 1858.

*To the Common Council of the City of New York:*

GENTLEMEN—I am not sure that you ought to have invited a Postmaster-General to be present at your great celebration. The event you commemorate most emphatically smashes up his old stages, runs his cars off the track, and plays the mischief with his post-office establishment generally.

I remember when the first steamboat made its appearance on one of our western rivers. Before that time, all our trade and commerce were with New Orleans, by flatboats, commonly called "Broad horns." When the magnificent steamer came dashing along and passed a fleet of the "broad horns" with a speed that seemed incomprehensible, the hands, dropping their oars and stretching their eyes wide with amazement, cried out "Farewell, broad horns, forever."

What must now be the exclamation of my poor stage-drivers, crawling along at the rate of four miles per hour, or of the little post-boy, bobbing along on his hard trotting horse, when he hears that the President and Queen Victoria have been exchanging congratulations between London and Washington in a single day. In all seriousness, if you New Yorkers shall continue in the notion of stretching a cable from your own great city, I shall be compelled to consider my Department as "seriously implicated."

The reflection has doubtless been consoling to my predecessors as well as to myself, that our Department was essential to all the great as well as small concerns of mankind, but now how sad the mistake: congratulations between mighty potentates, proclamations of peace and of war, orders between merchants and their distant factors, messages of undying love and devotion—true or pretended—all fly with the rapidity of lightning over our heads, looking down on our old fogy post-office arrangements with scorn and contempt.

But, gentlemen, I wish you to understand that I do not mean to surrender at discretion. One half of the news to be transmitted from one place to another is *bad news*, and nobody cares how slowly that goes; and so I hope, after all, I can give the telegraph a pretty hard scuffle for one half of the business, at least.

Pardon me, gentlemen, for this apparent trifling with a great subject. It is some little relief to the mind from that intense feeling of exultation and joy which success in laying the Cable has everywhere inspired. Time, space and the ocean have been subdued. Three great American names, Franklin, Morse and Field, stand identified with the noble work. May its blessings to mankind be commensurate with the grandeur of their achievement and the immortality of their fame.

Very hastily, your obedient servant,

AARON V. BROWN.

---

[From His Excellency N. P. BANKS, Governor of Massachusetts.]

COMMONWEALTH OF MASSACHUSETTS,
EXECUTIVE DEPARTMENT, COUNCIL CHAMBER,
BOSTON, Sept. 1, 1858.

GENTLEMEN—I have the honor to acknowledge the receipt of your kind invitation to a seat at the municipal dinner, given to Cyrus W. Field, Esq., and the officers of the ship Gorgon and the frigate Niagara, "in commemoration of the laying of the Atlantic Cable." Imperative official duties will deprive me of the pleasure of meeting you; but I cannot withhold an expression of my own grateful feelings, which I am sure will mingle with your more general joy, for the success of the great work you celebrate.

It is perhaps impossible for us exactly to measure the influences which instant communication between the European and American continents will work upon the industrial interests of our country. All the great inventions and discoveries of the last three centuries have contributed incalculably to the elevation of industrial pursuits. They have enlarged the resources of men and enabled them to supply their constantly increasing wants. Humanity keeps pace with material prosperity, and the development of great truths in the material world seems to contribute as directly to the equality of men and the welfare of the human family as the highest progression in institutions of government. As it has been in other cases it will be in this, the greatest of all. The power of an international and oceanic telegraph is so vast as to place it beyond monopoly of individuals or nations. The rich and poor, dependent colonies and controlling empires, must share alike, the world over, in its advantages. It will tend to the destruction of the selfish spirit of isolation in governments, which is barbarism, and to facilitate universal and friendly intercourse, which is civilization; to the recognition of the rights of all men, of the supremacy of intellect and truth, of the laws of justice, as well as to a just comprehension of the relations of men to each other and to God, which is religion.

I cannot wonder that New York, which must be, whether or no we acknowledge it, the commercial emporium of the New World, should hasten to give ocular and audible demonstration of her appreciation of this great achievement. I should with alacrity join you did not stern duty forbid; but I claim at your hands, before the last shout in honor of the triumph of energy and intellect is raised, a remembrance of the old Commonwealth that gave to the world FRANKLIN, MORSE and FIELD.

I am, most respectfully, your obedient servant,

NATHANIEL P. BANKS.

THOMAS McSPEDON, Esq.,

*And Gentlemen of the New York City Government.*

[From His Excellency E. DYER, Governor of Rhode Island.]

STATE OF RHODE ISLAND AND PROVIDENCE PLANTATIONS: EXECUTIVE DEPARTMENT.

PROVIDENCE, August 28, 1858.

GENTLEMEN--The Annual Commencement of our University, on the 1st of September (and at which the Executive of the State is expected to be present), and an engagement at a Sabbath School Anniversary on the following day, must deprive me of the pleasure of joining in the congratulations of my fellow-citizens upon the success of the Atlantic Cable, at the "municipal dinner" to which you have honored me by your invitation of the 25th inst.

If present, I should propose this sentiment: "The Atlantic Cable—In its annihilation of time and distance between the Old and New World, may it become the perpetual bond of universal peace and fraternity."

Very respectfully, yours,

ELISHA DYER.

To Messrs. THOS. MCSPEDON, *Chairman*, and C. T. MCCLENACHAN, *Sec'y, Board of Councilmen, New York.*

---

[From His Excellency S. P. CHASE, Governor of Ohio.]

STATE OF OHIO, EXECUTIVE DEPARTMENT.

COLUMBUS, August 31, 1858.

DEAR SIR—Nothing could be more gratifying to me than to participate with the citizens of New York in the celebration of the greatest event of our day. I deeply regret that it is impossible.

It well becomes the great commercial emporium of the Union to take the lead in the celebration of an event so auspicious to commerce. New York may be justly proud that the successful accomplishment of the great work is due mainly to the intelligent energy and unconquerable perseverance of one of her own merchants. As citizens of the same great republic, we, in Ohio, partake her joy and share her triumph. We rejoice that a MORSE prepared the means, that a MAURY explored the way, and that a FIELD achieved the work of uniting the Old World and the New by these wondrous electric sympathies. And we trust that what has been done may but prelude greater things to be, and that the vast scheme of intercontinental communication, inaugurated by prayer, and closed, in its first stage, by thanksgiving, never more appropriate than on the lips of the gallant HUDSON, may be consummated while we yet live, amid rejoicings, following the flash, and hasting before the sun, around the world. So mote it be.

With great respect, yours truly,

S. P. CHASE.

To THOMAS MCSPEDON, Esq., *Chairman Joint Committee of Common Council of New York.*

---

[From His Excellency A. P. WILLARD, Governor of Indiana.]

EXECUTIVE DEPARTMENT, INDIANA,
INDIANAPOLIS, August 28, 1858.

DEAR SIR—I received the invitation through you, of the Joint Committee of the Common Council of the city of

New York, to be present at the municipal dinner to be given on the 2d of September, to commemorate the successful laying of the Atlantic Cable.

I regret that circumstances forbid my attendance. But allow me to join with you in congratulating all, that every day of peace increases the success of science and enlarges the hope of the philanthropist, that all civilized and Christian people will at length find their greatest interest and happiness in cultivating friendly, financial and commercial relations.

Most respectfully,

ASHBEL P. WILLARD.

THOS. MCSPEDON,

*Chairman, New York.*

---

[From His Excellency R. M. STEWART, Governor of Missouri.]

EXECUTIVE DEPARTMENT,
CITY OF JEFFERSON, August 30, 1858.

C. T. MCCLENACHAN, Esq.:

DEAR SIR—Your letter, dated August 25th, written in behalf of the Joint Committee of the Common Council of the city of New York, inviting me to be present at the municipal dinner to be given on the evening of September 2d, at the Metropolitan Hotel, in celebration of the laying of the Atlantic Telegraphic Cable, has just come to hand.

I regret that official duties must necessarily prevent my attempting to complying with said invitation, even if it were possible to reach your city in season. Feeling a deep interest in whatever tends to characterize this as a truly progressive age, and ready on all occasions to encourage any enterprise calculated to facilitate the friendly intercourse of the human family, or to promote the great interests of commerce (with which all others are interwoven), the event you celebrate elicits my warmest sympathies, and were it in my power it would afford me great pleasure to participate in the festivities of the occasion.

The interoceanic telegraphic communication between the two great continents, is an epoch in the history of the world, which is fraught with results too great to be now conceived by the human mind.

Again expressing my regrets that it is impossible for me to be present at the festive board to which you have invited me, as the Chief Executive of Missouri—the great central State of the American Union, and ultimately, in my humble opinion, to be the great commercial heart of the American continent; and reassuring you that nothing would afford me greater pleasure than to join you in the celebration of an event which must so quicken the pulsations of the world of mind, and so connects the nervous texture of the great family of man that it must ultimately, and at a period not very remote, unite in one common brotherhood all the civilized nations of earth, and rapidly hasten the day when the arts and sciences shall be cultivated throughout the habitable globe, I beg to assure you, and those whom you represent, of my high esteem.

Very truly, yours,

R. M. STEWART.

[From His Excellency H. H. SIBLEY, Governor of Minnesota.]

STATE OF MINNESOTA, EXECUTIVE OFFICE, SAINT PAUL,

August 31, 1858.

THOMAS McSPEDON, Esq.,

*Chairman Joint Com. Common Council, N. Y. City.*

DEAR SIR—I have the honor to acknowledge receipt of your favor of the 25th instant, in which the congratulations of the Joint Committee are extended to the Governor of Minnesota, on the successful laying of the Atlantic Telegraph Cable, and requesting him to attend the municipal dinner, to be given at the Metropolitan Hotel, in New York city, on 2d September next.

It being impossible for me to participate in your festivities, it only remains for me to express my obligations for your courtesy, and to assure you that the event you are about to celebrate in your great city is hailed with as much interest and delight by the people of Minnesota as by those of any portion of the country. Our capital, St. Paul, will to-morrow be the scene of ceremonies and rejoicing in honor of the successful union of Europe and America by telegraph. I tender you my thanks for your generous invitation, and, on the part of our youthful State, I cordially reciprocate the congratulations you have been kind enough to tender to her Chief Magistrate, on her behalf.

I have the honor to be,

Very respectfully,

Your obedient servant,

HENRY H. SIBLEY.

[From His Excellency THOMAS BRAGG, Governor of North Carolina.]

RALEIGH, August 28, 1858.

MY DEAR SIR—I had the honor to receive, last evening, your invitation, in behalf of the Joint Committee of the Common Council of the city of New York on celebrating the laying of the Atlantic Cable, to the municipal dinner at the Metropolitan Hotel, on the evening of the 2d of September next.

It would be very gratifying to me to visit your great city on an occasion so interesting, and especially to unite with her citizens in the celebration of an event which must hereafter be marked as an epoch in the history of the world. My official engagements are, however, such that I cannot conveniently be absent, and I therefore, very reluctantly, decline the invitation.

With sincere thanks for the attention,

I am,

Most respectfully and truly, yours,

THOMAS BRAGG

THOMAS McSPEDON, Esq.

---

[From His Excellency R. F. W. ALLSTON, Governor of South Carolina.]

EXECUTIVE DEPARTMENT, GLENN SPRINGS,
September 3, 1858.

The Governor of South Carolina promptly reciprocates the sentiment of congratulation conveyed in the polite invitation of the Joint Committee of the Common Council of

New York, on the occasion of the completion of the Atlantic Telegraph, the latest and noblest application of Morse's happy invention. May this great work insure "peace on earth and good-will to man" from his brother. Its success, so far, is unmistakable evidence of good-will from our common Father.

As the great centre of commerce in America will be benefited in the greatest degree by the new and wonderful facility of inter-communication with Europe, so will her responsibility be increased to preserve that peace and good-will, to furnish to the world examples only of integrity and conscientious dealing. "To whom much is given, of him shall much be required."

Having been engaged for a month in a military tour through the Fifth Division of the State, your letter did not reach the Governor until yesterday.

It is not doubted that the celebration has been one worthy of all concerned in the stupendous effort—alike honorable and satisfactory to the heart of the financial system of America which animates you.

Very respectfully,

R. F. W. ALLSTON.

*To the Joint Committee on Celebration of the Common Council of the City of New York.*

---

[From his Excellency Sir EDMUND HEAD, Governor of Canada.]

TORONTO, August 23, 1858.

To THOMAS MCSPEDON, *Chairman, &c., City of New York:*

I beg to thank the Committee of the Council of New

York for their courteous invitation, and regret that public duty prevents me accepting of it.

EDMUND HEAD.

---

[From His Excellency A. BANNERMAN, Governor of Newfoundland.]

ST. JOHN'S, N. F., August 28, 1858.

*Chairman of the Committee on Celebration, New York:*

SIR—The Governor of Newfoundland feels much gratified by the invitation with which he has been honored by the Common Council of New York, to be present at the festival to be held in that city, early in September, to celebrate an event which has been brought to a successful and happy issue by the united efforts of American citizens and British subjects.

Although the Governor is not allowed to leave the Colony without permission from home, on an occasion like this he would have visited New York at all hazards, but the Admiral Commander on this station having intimated a visit here, the Governor can only return his thanks to the Common Council of New York, and beg they will accept the assurance of his regards and sincere wishes that our respective nations may ever continue to live in the bonds of good-will, friendship and peace.

A. BANNERMAN,

*Governor, &c.*

---

[From His Worship the MAYOR of QUEBEC.]

CITY HALL, QUEBEC,

30th August, 1858.

The Mayor of the City of Quebec regrets that official

business prevents his being able to comply with the kind invitation of His Worship the Mayor and the Common Council of the city of New York, to be present at the parade and festivities to take place at the Crystal Palace, and at the municipal dinner to be given at the Metropolitan Hotel.

The Mayor of the city of Quebec takes this opportunity of accepting the congratulations of His Worship the Mayor of the city of New York on the occasion of the successful laying of the Atlantic Cable, and hopes that that metallic cable will unite more and more two great people whose common origin seems to have prepared for them one common life one common glory.

HECTOR L. LANGAIM,

*Mayor.*

---

[From His Worship the MAYOR of HAMILTON, C. W.]

HAMILTON, 30th August, 1858.

To THOMAS MCSPEDON, Esq., *Chairman of the Joint Committee of Common Council on Celebration, New York:*

SIR—I have the honor to acknowledge the congratulations of the Mayor and Common Council of the city of New York, occasioned by the successful completion of the Atlantic Telegraph, also the receipt of their kind invitation to attend the celebration of the greatest event of the age; and I beg to say, in reply for the Common Council of the city of Hamilton and myself, that they view with the liveliest interest the consummation of a work which stands in the triumphs of science.

I sincerely regret my inability to become a participant in the festivities of the occasion.

I have the honor to remain, sir,

Yours, obediently,

GEO. H. MILLS,

*Mayor of Hamilton.*

---

[From Rear-Admiral Sir HOUSTON STEWART, R. N.]

HALIFAX, August 14, 1858.

C. T. MCCLENACHAN, *Secretary, &c.:*

SIR—Your telegrams of yesterday have both reached me. I am much flattered by the invitation of the Common Council of the city of New York to be present at the celebration of that great and most important achievement which has just been accomplished, and it is with much regret that I find myself compelled to decline the distinguished honor, my duties demanding my presence elsewhere.

With respect to the Gorgon, I have already signified to Mr. Field my inability to interfere, in any way, with Captain Dayman's proceedings. He is not under my orders. I have no notice whatever concerning him, and he has instructions direct from the Admiralty, by which he must be guided.

I beg you will convey my grateful acknowledgments to all the members of the Common Council, together with my cordial good wishes for the successful issue of their laudable intention to commemorate an event which, I earnestly hope, will materially tend to cement those feelings of amity and good-will which ought ever to prevail between the United States and Great Britain.

HOUSTON STEWART.

[From His Excellency BARON OSTEN SACKEN, of Russia.]

LEGATION OF RUSSIA,
WASHINGTON, August 30, 1858.

*To the Mayor and Common Council of the city of New York:*

GENTLEMEN—His Excellency Mr. Stoeckl having sailed lately for Europe, I will not fail to inform him of your kind invitation to assist at the celebration of the successful laying of the Telegraph Cable.

I feel confident that, although absent, Mr. Stoeckl will feel grateful for this attention of the city authorities of New York, and will heartily join them in their wishes that this great enterprise, now successfully achieved, might prove useful towards the promotion of the common welfare and the peace of nations.

I have the honor to be,

Your most obedient servant,

BARON OSTEN SACKEN,

*Chargé d'Affaires.*

---

[From His Excellency COUNT SARTIGES, Chargé d'Affaires of France.]

NEWPORT, R. I., September 1, 1858.

*To the Common Council of the city of New York:*

I regret very much that a severe indisposition prevents me from accepting your invitation to dinner for to-morrow evening.

I beg you to accept my most sincere congratulations upon the successful laying of the Transatlantic Cable. I may add, with confidence, the most cordial congratulations of the Emperor and of the whole French nation upon this new binding of Europe to America.

COUNT SARTIGES.

---

[From the Austrian Consul General, C. F. LOOSEY, Esq.]

K. K. ŒSTERREICHISCHES GENERAL CONSULAT,
NEW YORK, 2d September, 1858.

GENTLEMEN—I have the honor to acknowledge the receipt of your highly esteemed invitation to attend upon the celebration of the laying of the Atlantic Cable, and regret exceedingly that indisposition prevents me from assisting at this memorable event, which, by the blessings it is destined to confer upon mankind, will stand out in bold relief in the annals of the history of the world for ages to come.

I beg you to accept my most sincere thanks for your kind attention, and, reiterating my expressions of regret, I have the honor to subscribe myself, gentlemen,

Your most obedient servant,

CHARLES F. LOOSEY,

*Austrian Consul-General.*

THOMAS MCSPEDON, Esq.,

*Chairman of the Joint Committee of the Common Council.*

C. T. MCCLENACHAN,

*Secretary.*

---

[From His Excellency G. G. TASSARA, Minister from Spain.]

NEW YORK, September 2, 1858.

SIR—I came to this city with the earnest desire of joining in the celebration of the successful laying of the Atlantic Cable, to which I had the honor to be invited; but, to my regret, an indisposition prevents me from being present at the municipal dinner that is to take place to-day at the Metropolitan Hotel, and I am compelled to beg of you the favor of presenting my apologies and acknowledgments to the Honorable the Common Council of this city, together with the assurances of my heartfelt concurrence in the sentiments which have prompted the commemoration of this very great event.

I am, sir, most respectfully,

Your obedient servant,

GABRIEL G. TASSARA.

To THOMAS MCSPEDON, Esq., *Chairman, &c.*

---

[From the Spanish Consul, S. STOUGHTON, Esq.]

NEW YORK, 2d September, 1858.

*To the Honorable the Common Council of the City of New York, &c.:*

The Spanish Consul begs to offer his sincere thanks to the Honorable the Common Council of the city of New York, for the honor of their invitation to join them to-day at the municipal dinner, to be given by them to Cyrus W. Field, Esq., the officers of H. B. Majesty's Ship Gorgon, and U. S. Steam Frigate Niagara, in commemoration of the laying of the Cable, and begs to offer his regrets that, owing to his continued indisposition—occasioned by a severe fall—he will be unable to participate in their hospitality and render personally to their distinguished guests that homage and congratulation due to merit.

[From the Rt. Reverend JOHN J. MULLOCK, Bishop of Newfoundland.]

ST. JOHN'S, September 1, 1858.

*To the Common Council of the city of New York:*

GENTLEMEN—I feel most grateful to you for the very high honor you have done me by inviting me to participate in the festivities of the great city of New York, on the happy completion of the most wonderful enterprise of the age—the Atlantic Telegraph. Nothing but the want of due means of getting to New York in time, would prevent me of availing myself of your honorable invitation, but no steamer leaves this in time for the 1st or 2d of September. Though absent, however, I join with you in spirit, and pray God that the union now made between the two greatest and kindred nations of the world may be a perpetual bond of peace, and that the only messages may be that of common and mutual interest. While in other lands millions of armed men are sternly watching each other's frontiers, a slender wire will be the great link of union, peace and good-will between the two great and kindred nations of Britons and Americans.

I have the honor to remain, gentlemen, your most grateful, humble servant,

†JOHN J. MULLOCK,
*Bishop of St. John's, Newfoundland.*

THOMAS MCSPEDON, Esq.,
*Chairman Common Council Committee, New York.*

---

[From HON. EDWARD EVERETT.]

BOSTON, August 30, 1858.

DEAR SIR—I have received your obliging letter of the

26th, inviting me to attend the dinner of the 2d of September, and to respond to one of the regular toasts.

I need not say that I fully sympathize with you on this great occasion, and that I should be most happy to join you in doing deserved honor to Mr. Field, and to all others connected with the all-important achievement which you propose to celebrate. I much regret that official duties, as a Trustee of the Public Library, require me to be in Boston on Thursday.

With respect to the event itself, the imagination sinks under the effort to measure the probable results of the communication of thought with electric speed throughout the civilized world. We must remember, too, that, great as is the progress which has been made, we are yet in the infancy of this system of communication. The next link in the wondrous chain, at least in this part of the world, will be to connect the Pacific with the Atlantic coast. Auxiliary lines will probably, in the course of this generation, be extended through Oregon, Washington Territory and New Columbia to the Russian settlements on the Continent. Asia is separated but by a narrow strait; and if a more southern route than that across Behring's strait is desirable, it is furnished by the Aleutian Islands, with very moderate oceanic intervals, to the mouth of the Amoor River, whose connection with the more western portions of the Russian Empire has lately been traced by our enterprising countryman, Mr. Collins.

The lines already established connect not merely New York, Philadelphia and Boston, with Liverpool and London, but they connect every State and every city of our own vast continent with every kingdom and every city of England and Continental Europe.

May they prove, under Providence, the efficient means, not merely of confirming our amicable relations with foreign countries, but of strengthening the bonds of good feeling and patriotic affection with every portion of our own.

With the best wishes for a successful festival, I remain, dear sir, your friend and fellow-citizen,

EDWARD EVERETT.

THOMAS McSPEDON, Esq.

---

[From Hon. L. TREMAINE, Attorney-General of the State of New York.]

STATE OF NEW YORK,
OFFICE OF THE ATTORNEY-GENERAL,

ALBANY, August 31, 1858.

DEAR SIR—I have the honor to acknowledge the receipt of an invitation to attend the municipal dinner, in commemoration of the laying of the Atlantic Cable, and regret that other engagements will deprive me of the pleasure of accepting it.

The great event you celebrate will form a memorable era in the history of the world. It is too early to comprehend all the vast results which are destined to flow from it; but as indicating progress towards the period when the "lion and the lamb shall lie down together," it may be hailed with lively satisfaction by the friends of humanity.

With my best wishes for your success, I am

Yours, very truly,

LYMAN TREMAINE.

T. McSPEDON, Esq.,

*Chairman, &c.*

[From Captain THORBURN, U. S. S. Sabine.]

U. S. SHIP SABINE, OFF THE BATTERY,
NEW YORK, September 1, 1858.

C. T. MCCLENACHAN, Esq., *Secretary, &c.*

SIR—I have the honor to acknowledge the receipt of an invitation, from the Common Council of the city of New York, to be present at a municipal dinner to be given on the second instant to the gentlemen who have conducted to a successful issue the laying of the Atlantic Telegraph Cable, commemorative of that great achievement.

I will ask you to offer to the Council my heartfelt thanks for the honor which they have conferred, in asking me to participate upon this great congratulatory occasion, and my extreme regret that present indisposition constrains me to decline.

In declining I would also express to them my high appreciation of the event to be celebrated, and the entire sympathy with the feeling which prompts its commemoration. We justly hail it as the most important step in the march of progress which the world has ever known, and as justly hope that this chain which unites the two continents, spread through the depths of that ocean upon whose bosom their hostile navies have met, shall prove a bond of perpetual peace, and shall commence an epoch of eternal unity.

To join with the demonstrations on shore to-day, as this ship was passing from the Navy Yard to her present anchorage off the Battery, the ensign of our kindred nation was hoisted at the fore and our own at the main and mizen. After the ship was moored, a salute of thirty-two guns was fired.

In conclusion, let me again, through you, express my high sense of the compliment paid me, and sincere sorrow at not being able to be present to accord my thanks by word of mouth.

I am, respectfully, your obedient servant,

ROBERT D. THORBURN,

*Commanding U. S. S. Sabine.*

---

[From Professor JOSEPH HENRY, of the Smithsonian Institute.]

SMITHSONIAN INSTITUTE,
WASHINGTON, D. C., August 31, 1858.

*To the Committee of the Common Council of the City of New York:*

GENTLEMEN — I write to express my thanks for the honor you have conferred upon me in the invitation to attend the ceremonies and dinner to celebrate the laying of the Atlantic Cable, and to compliment Cyrus W. Field, Esq., and the officers of H. B. M. steamship Gorgon and the U. S. steam frigate Niagara. If other engagements did not absolutely forbid my acceptance it would give me great pleasure to be present on this interesting occasion, and to partake of the sympathetic enthusiasm which must be awakened by communion with the men who have been so successful in accomplishing this latest step in the application of electricity to the affairs of life.

This is a celebration such as the world has never before witnessed. It is not alone to commemorate the achievements of individuals, or even of nations, but to mark an epoch in the advancement of our common humanity.

The Atlantic Cable does not merely connect, in sympathetic bonds of Christian fellowship, England and America, but its success insures the extension of similar inter-lines of thought between every part of the civilized world. The undertaking was fraught with responsibilities and difficulties of no ordinary kind. Had it failed, the loss would not have been merely the money and time spent in the attempt, but far more, the loss of confidence in the public mind as to the practicability of the enterprise, and the consequent impossibility for years to come of obtaining the means for another experiment. The difficulties were such as could only be appreciated by those who were well versed in practical engineering and in the theoretical principles of electricity. I need not say how fearlessly these difficulties were encountered by Mr. Field and his associates, and with what sagacity, forethought and perseverance they have been overcome.

The distinctive feature of the history of the Nineteenth Century is the application of abstract science to the useful arts, and the subjection of the innate powers of the material world to the control of the intellect as the obedient slaves of civilized man. To secure a result o the kind which we are now called to celebrate it was necessary, first, that scientific discovery should furnish the principles on which the invention was founded; secondly that ingenuity should supply the machinery and various appliances necessary to the accomplishment of the design thirdly, that men of enterprise, of confidence in their own ability and faith in the knowledge and resources of the times, should be found to embark in the undertaking.

Nor was it less indispensable that the public mind should be so impressed with the importance of the object that individuals would risk their fortunes in furnishing the necessary means, and Governments give it their protection and support. Such a concurrence of circumstances could not have happened at any other period of the history of the world, and emphatically distinguishes the event as belonging to the middle of the present century.

While I put full faith in the advance of civilization, principally through the application of science, of education and the cultivation of the moral faculties, I do not believe that this result is to be produced by a blind law of destiny, but providentially through the intervention of individual agencies. The discoveries of every scientist, from Thales, of Miletus, to those of Faraday and Thompson, have added to our powers over electricity, that most subtle of the multiform agents of nature, and each distinguished inventor, including Morse, House, Hughes and others, has assisted in producing the result which to-day excites our admiration and calls forth our gratitude—a result which gives bright hopes for the future and points to conquests yet to be achieved; which promises (may it not be hoped?) peace on earth, good-will to men and glory to God in the highest.

I have the honor to be,

Very respectfully,

Your obedient servant,

JOSEPH HENRY.

Grace was said by the Rev. Dr. Ogilby, assistant minister of Trinity Church.

After dessert, the company was called to order by the Mayor, and he then announced the first regular toast—"THE PRESIDENT OF THE UNITED STATES."

The toast was duly honored; the company standing, while Dodworth's band played "Hail Columbia," and then giving three cheers for the President, when the flag, which was waving over his portrait, was withdrawn.

The second regular toast was then given—"THE QUEEN OF GREAT BRITAIN AND IRELAND." This toast was honored with the greatest enthusiasm; the band playing "God save the Queen," and the company standing and saluting her Majesty's portrait with continuous cheering.

The third regular toast was then announced—"THE GOVERNMENT AND PEOPLE OF GREAT BRITAIN AND IRELAND—joined to us in the Court of Neptune, may the nuptials never be put asunder." The toast and sentiment having been duly honored, Lord Napier, her Majesty's representative at Washington, was presented to the company by the Mayor and was received with every demonstration of respect and honor. He responded to the toast as follows:

*Mr. Mayor and Gentlemen:*

When I received Her Majesty's orders to proceed to the United States, I flattered myself that I entered upon my duties at an auspicious time, and I cherished a hope that the period of my residence might be coincident with that solid and hearty reconciliation of our respective countries which the tendencies of the age transparently indicated to be near at hand. Nor have I been disappointed. The course of political affairs since my arrival has, indeed, exhibited some asperities which it was impossible to foresee, and which could not be regarded without concern; but, gentlemen, I now hail in the event which we are met to celebrate, a glorious compensation for past anxieties and an important security against future dangers. To be the contemporary and spectator of this great monument in human progress is alone a cause of honor and exultation. The triumph in which your distinguished guests have so high a share does not only confer on them the celebrity and affection which mankind bestow on their purest benefactors—such a triumph gives to the protecting Governments a claim to public gratitude and historic commendation, it adorns and dignifies the nations and the time in which it is wrought, it imparts the generous contagion of enterprise, it teaches the universal lesson of faith, patience, and perseverance, it infuses into men's souls a sense of conscious worth and pours on all, however humble and remote, the glow of reflected fame. I question whether any single achievement had ever more features of interest and utility; all the elements of adventure, difficulty and hazard have been here assembled which could arrest the imagination, and no virtue has been wanting which could satisfy the judgment and captivate the heart. Long will those scenes remain dear to the popular memory. With what admiration do we contemplate the cheerful enthusiasm of Field, inspiring shareholders and admiralties with kindred zeal, undiscouraged by the hostility of nature and the powers of science, divining success where others denounced defeat, and carrying off the palm before an incredulous though sympathizing world. What encounter on the sea can compare with that last meeting of the confederate ships when the knot was knit which shall never be rent asunder? How anxiously do we follow the Gorgon on her constant course, and watch the Niagara threading the icebergs and traversing the gloom upon her consort's helpful track. We tremble for the over-burdened Agamemnon, still tormented by the gale. We blend our aspirations with the worthier prayers of Hudson, when he kneels, like Columbus on the shore, and invokes the Divine protection on his accomplished work. Nor is the moral aspect of this great action marred by any mean infirmity. Here there is nothing to obliterate, nothing to deplore. The conduct of the agents exemplified the purposes of the deed; with manly emulation, but inviolate concord, they cast forth upon the waters the instrument and the symbol of our future harmony. This is not the place to demonstrate the usefulness of telegraphic communication in the practice of government and commerce, and its numerous consolations in matters of private affection. I content myself with recognizing its value in international transactions. Something may be detracted from the functions of diplomacy, but much will undoubtedly be gained for the peace of nations. By this means the highest intelligence and authority on either side will be brought into immediate contact, and whatever errors belong to the employment of subordinate and delegated agencies may be prevented or promptly corrected. By this means many of the evils incidental to uncertainty and delay may be canceled, offenses may be instantly disavowed, omissions may be remedied, misapprehension may be explained, and in matters of unavoidable controversy we may be spared the exasperating effects of discussion proceeding on an imperfect knowledge of facts and motives. In addition to these specific safeguards it may be hoped that the mere habit of rapid and intimate intercourse will greatly conduce to the promotion of a good understanding. On the one hand stands England, the most opulent and vigorous of monarchies, in whose scant but incomparable coil lie compacted the materials of a boundless industry, on the other the Republic of the United States, founded by the same race, fired by the same ambition, whose increase defies comparison and whose destinies will baffle prediction itself. We cannot doubt that these fraternal communities are fated to enjoy an immense expansion of mutual life. The instant interchange of opinion, intelligence and commodities, will become a condition almost inseparable from existence; and whatever stimulates this development will oppose a powerful obstacle to the rupture of pacific relations. No man of common liberality and penetration will question the position and certain merits of a discovery which has connected England with America, and America with the whole civilized world besides. I would not darken the legitimate satisfaction of the present moment by uttering a reluctant or sceptical estimate of our new faculty. Yet, even in this hour of careless and convivial felicitation we shall do well to remember that the magnetic telegraph forms no exception to the category of inventions which, however apt and proper and willing to be the vehicles of benevolent designs, are also the unresisting tools of every blind or intemperate impulse in our nature. The votaries of a querulous philosophy speciously assert the unequal march of morality and mind; and even a poet has affirmed in foreboding verse, that all the train of arts which have reduced the material elements to be the vassels of our will

Heal not a passion or a pang
Entailed on human hearts.

It belongs to our respective countries and to the present age to confound that speculation which would divide knowledge from virtue, and inquiry from improvement. The labor will not be light, nor is the eventual victory everywhere apparent, yet there is one province of affairs in which the task would be easy and the triumph within our grasp. It depends on us, on our will, on our choice, to carry into perpetual effect the sentiment which the honorable chairman has associated with his toast; it depends on us to strike out forever from the sum of public and social embarassments all the contingencies of a collision between England and America. If we should not employ our unprecedented powers in a friendly spirit, if we should hereafter offer unreflecting provocation and conceive hasty resentment, if every transient cloud which ascends on the political sky be hailed as the prognostic of a destructive storm, if we should make haste to unlock the well of bitter waters and to raise the phantoms of extinct pretensions and buried wrongs, then would this memorable effort of ingenuity and toil be partly cast away. Gentlemen, I am confident that we shall pursue a very different course. The Queen has sent tidings of good-will to the President, and the President has made a corresponding answer to the Queen. Those messages must not be dead inscriptions in our archives, they must be fruitful maxims in our hearts. Let our Government be considerate in their resolutions. Let the orators of one country comment upon the institutions, the policy, and the tendencies of the other in a candid and gentle spirit. Let the negotiators of both approach the adjustment of disputed questions, not with a tenacious regard to paltry advantages, but with a broad view of general and beneficial results. Then, gentlemen, the subtle forces of nature will not have been employed in vain, and we shall give a worthy office to those subjugated and ministering powers which, by divine permission, fly and labor at our command.

Mr. Mayor and gentlemen, the manifestation of respect for the Queen which you have given to-night, and which has been apparent throughout these celebrations, will be highly appreciated by Her Majesty and by her faithful subjects, who observe with pride that the virtues of their sovereign have won back the spontaneous homage of a free nation. The Ministers of Great Britain will correctly estimate the momentous import of an enterprise to which they gave an effective support, and will, I am well assured, transport into our official relations the cordial sentiments which animate the English people towards their American kindred. I tender you my sincere thanks for the honorable welcome granted to my countrymen and myself. You have conferred on us a favor which we shall ever acknowledge, for your goodness has enabled us to associate our names and voices, however feebly and far off, with an event which must have an everlasting and benignant significance. We are all firmly persuaded that there exists here a deep and warm attachment to the mother-country, gathering strength with time and rejoicing to obtain a commensurate return. As the grateful, though inadequate, representatives of the British empire, we declare that the hands that are joined to-day are joined in sincerity, and the grasp that we have felt we desire to be eternal.

The fourth toast was—" OUR SISTER STATES AND SISTER CITIES: New York greets them, and trusts that the cord which binds the New World with the Old, welds more firmly the links of the Union."

The Mayor called upon Rodman M. Price, ex-Governor of the State of New Jersey, to respond to this toast. On rising to do so Mr. Price was loudly cheered. He said:

*Mr. President and Gentlemen:*

At being thus suddenly called upon to respond to so gracious, so comprehensive and so patriotic a toast, I feel very much as if I should rather adopt the admonition that I overheard in the crowd in Broadway yesterday, as given by one son of Erin to another, when they were beholding the imposing procession. Said one to the other

"Jimmy, d'ye think they're havin' as good time in London to-day as we are?"

"Divil a bit, Pat, they havn't the ability to get up any such celebration or procession as this."

"Be quiet, man," says Jimmy, "be respectful, be aisy, say nothin', or they'll telegraph what ye'll say to London!"

Mr. President, who can doubt that our sister States and sister cities cordially, enthusiastically and rapturously unite all their sympathies with you and us on this auspicious and interesting occasion. The greetings which you have already received from the authorities of our sister States and our sister cities show how intensely they feel the triumph which Science has won by the successful laying of the Atlantic Telegraph Cable — how much they rejoice that the Old World and the New are connected by a Transatlantic Telegraph Cable.

You must not expect me, sir, to answer for all the States or the cities of this glorious Union. It is impossible for me to express the feelings of those remote bounds of our Union, as well as the hundreds of densely populated cities. They have spoken, sir. I have seen their congratulations, transmitted through the telegraphic wires, published, and we know that their feelings are as intense as ours. How could you expect me to respond for the first colony of Massachusetts, or for the enterprise of State street, Boston? How could you expect me to respond for the glorious Keystone State Pennsylvania, or for her Empire City of Philadelphia, or for Maryland, or her Monumental City? Much less can I be expected to respond for the Old Dominion. Neither am I in the temper to respond for the Crescent City of the South, or for the Queen City of the West, or for all the lesser cities to the extreme point of our territories.

No, sir. I will not occupy the time of this company in speaking of localities. It is enough to know, sir, that they rejoice with us upon this great and glorious occasion — that they exult with us in this triumph of Science. That great procession of yesterday will be a recorded memory to the end of time. Sir, though this city has done nobly, and while I speak in behalf of your sister States and cities, I must not forget my own glorious little revolutionary State, New Jersey. I thank you, sir, and the Common Council of this great city, for the recognition they have made of New Jersey as a sister State.

If I may be allowed to express some of the present pleasures and gratifications which I feel upon this occasion, I must allude to the great gratification which I have felt in the fact that our glorious navy has one peaceful triumph, and that the wooden bulwarks of England share with them the glory. My early association with the Navy of the United States makes the distinction which that arm of our service has won particularly agreeable to me, for I know, personally, and have been associated with some of the officers of the Niagara, and I can answer that no more devoted or zealous men wear the button and uniform of the American navy.

To our English friends who are present upon this interesting occasion, permit me to say that I reciprocate most cordially the noble sentiment of the poet Mackay, in a delightful poem which he wrote upon American soil during his late visit to this country, called "Brother Jonathan to Brother John," in which he makes Brother John utter the following sentiment to Brother Jonathan :

Said Brother John to Jonathan,
"You take the West and I the East;
We speak the self-same tongue
That Byron wrote and Chatham spoke,
And Burns and Shakspeare sung."

Sir, the sentiment of this toast trusts that the chord which binds the New World with the Old will weld more firmly the links of this glorious Union. Who can doubt, sir, that this Cable will prove a cord that will make the bonds of our sister States and our sister cities more indissoluble, and while we trust that that bond will unite us more cordially with the Old World, we feel and know that it will prove a bond of harmony, concord and peace with our sister States and sister cities.

Mr. Price was frequently applauded throughout, and sat down amid great cheering.

The fifth toast was then announced, and was drunk with all the honors.

"CYRUS W. FIELD—To his exertions, energy, courage and perseverance are we indebted for the Ocean Telegraph; *we* claim, but *immortality* owns, him."

Mr. Cyrus W. Field responded. He said:

*Mr. President and Gentlemen:*

You do me too much honor. The unparalleled celebration of yesterday and this banquet, dazzling as they have been, have not so far blinded me that I do not know them to be intended for more than myself. To no one man is the world indebted for this achievement. One may have done more than another—this person may have a prominent, and that a secondary, part; but there is a host of us who have been engaged in the work, the completion of which you celebrate to-day. I will mention my brother Dudley, with whom, as well as myself, the scheme originated, and who has been my adviser and helper throughout; my excellent friends, Messrs. Peter Cooper, M. Taylor, M. O. Roberts and W. G. Hunt, who have spent their money so generously and borne up so manfully under great discouragements; Mr. Brett, Mr. Whitehouse and Mr. Bright, with whom I made an agreement in September, 1856, under which the Atlantic Telegraph Company was formed, after I had obtained the authority of the New York, Newfoundland and London Telegraph Company, to make arrangements in England for the completion of their line across the Atlantic, either by a new subscription to their Company, or by a new Company; the British Government that came forward so nobly, in connection with which are especially to be mentioned Lord Clarendon and Sir John Packington, the present, and Sir Charles Wood, the late First Lord of the Admiralty; as also the Secretary of the English Treasury and Captain Washington, hydrographer of the English Navy; the Government of the United States—and I would particularly mention the President and Secretary of the Navy—the Governments of Newfoundland, Prince Edward Island, Nova Scotia, Canada and Maine; those never-to-be forgotten philosophers Lieutenant Maury, Professor Morse, Professor Farrady, Professor Bache and Professor W. Thompson, who have rendered most efficient aid without receiving any compensation; the accomplished and skillful English commanders Preedy, Dayman, Aldham, Otter, Wainright and Noddall; our own well known and honored Captains Hudson and Sands, Lieutenant Berryman and all the gallant officers of the Niagara, and the whole telegraph fleet, never forgetting the brave English and American sailors; Mr. William E. Everett, the able engineer who has devoted his whole time for months to constructing and perfecting the paying-out machinery with which the Cable has been so successfully laid; Mr. Woodhouse and Mr. Canning, engineers; Mr. Appold, who has been unwearied in his exertions on our behalf; the consulting committee of Engineers, Messrs. Joshua Field, John Penn, and Thomas Lloyd, who, without compensation, gave the Company the benefit of their eminent talent; the Directors of the Atlantic Telegraph Company, who have devoted so much time without receiving one shilling for their services, and my consistent faithful friends, Messrs. C. M. Lampson, George Peabody, and Junius S. Morgan, of London. I might add many—many others to this list. We have all worked faithfully and as efficiently as we could in the prosecution of the good work. We have had many difficulties to surmount, many discouragements to bear, and some enemies to overcome, whose very opposition has stimulated us to greater exertion. We believe that the work accomplished will prove a great and lasting benefit to our race. If it effect but a tithe of what is expected from it, we shall be more, much more than rewarded for what we have done.

The sixth toast was: "THE NAVIES OF GREAT BRITAIN AND THE UNITED STATES—Met and joined in a noble work of Peace; may they never be separated or meet in strife."

Captain Dayman, of H. B. M. steamship Gorgon, was called upon by the Mayor to respond to this toast, and did so in these terms:

*Mr. Mayor and Gentlemen:*

I feel highly honored in responding for the British navy to the toast you have just done me the honor of drinking, and I am desired by the distinguished Admiral, the Commander-in-Chief on the North American station, Sir Houston Stewart, to express to you his regret that he cannot himself be present at this celebration; but he has permitted several of the officers of his flagship to accompany us, and they are now here. It does not often fall to the lot of a naval officer, whose profession is chiefly that of war, to be associated with the national ships of another Power in duties strictly pacific. But in the service whose completion we are now celebrating, the United States ships Susquehanna and Niagara were associated and worked together with no less than six of Her Majesty's ships, viz., the Agamemnon, Leopard, Cyclops, Valorous, Gorgon and Porcupine. They co-operated heartily and successfully, to the mutual benefit of both navies, by the free and unconstrained intercourse commenced in the beginning and kept up to the end; and I trust that the rivalry thus engendered may long continue to be the only source of difference between us. It is, perhaps, somewhat presumptuous in me to speak of an officer so much my senior as Captain Hudson; but I cannot help saying, gentlemen, that Captain Hudson gained in England the friendship and esteem of officers of the highest rank in the navy, and that he carried away with him the best wishes of all who had the honor of his acquaintance. We witnessed a spectacle yesterday which will not be easily forgotten by any one of us. It was that of the enthusiastic rejoicings of tens of thousands of the freemen of this magnificent city, which stands first of all the cities of the American continent—rejoicings for a peaceful triumph, accomplished by the united efforts of Americans and Englishmen. We saw, as in our own happy country, multitudes congregated, without coercion or force to preserve good order. We saw, also, your volunteer regiments march past, and were struck with their steadiness and military bearing. Lastly, we joined in the great procession of firemen, which, as a body of organized men, trained for the performance of a most difficult duty, requiring a combination and discipline in the midst of destruction and confusion, appear to be the grandest and the most perfect institution of the kind in the world. And now, gentlemen, in behalf of the officers of the Indus and Gorgon, I beg to offer our warmest thanks for the hearty welcome which you have given us. We shall long remember with pleasure this our visit to your city, and shall carry back to England the most pleasing impressions of the feeling, consideration and kindness which has been shown us, and which we accept as indicative of that favorable regard for the old country which still lingers among her descendants on this vast continent.

The next toast in order was:

"THE ENGINEERS AND ELECTRICIANS who have done their work so well—The praise of both hemispheres shall be their reward."

In the absence of Gov. Seward, expected to answer to this toast, Rev. Dr. Bellows was called upon for a reply. Mounting the chair on which he had been sitting, he said:

*Mr. Mayor and Gentlemen:*

Being a poor speaker at the best, and now under the special disqualification of a short notice, I am determined to get whatever advantage is to be derived from a little artificial elevation. I certainly should not consent to speak for the Engineers and Electricians if this were not a telegraphic occasion, when everybody is subject to the law of telegraphic dispatch, and must be ready to take the place of any other gentleman, who, though a thousand miles off, is yet confidently expected—but has failed to appear—my precise predicament at this moment. If, through ignorance and inopportunity for preparation, I shall have little to say about the Electricians and Engineers—a topic which worthily to treat demands exact knowledge—you fortunately need few words to exalt your sense of their merit in this particular case. And, indeed, if we had our own elder Everett here, that matchless orator, he could not speak for the Engineers and Electricians with more effect than the modest silence of our younger Everett, whom we have here, has already done. The Engineers and Electricians of our country are everywhere speaking for themselves in these the days of their glory. Beginning with Franklin, who, let me say to the Committee of Arrangements, they will never be able permanently to *corner*, notwithstanding their success in placing him in an angle for this brief occasion (alluding to a scroll bearing his name, placed in a remote corner of the room), and coming through Fitch and Fulton, to Henry and Morse, it would be wholly beyond my power to add the least lustre to their reputations. As for the Company's "Woodhouse," it supplies the fuel for its own fame, while its "Bright" may be safely left to its own brightness.

Forsaking my text, so far as the Engineers and Electricians are personally concerned, I confess that in a more general aspect this toast has peculiar fascinations for me; for it is the first opportunity offered this evening to get off mere international grounds on to world-wide territory. It names both the hemispheres, and includes the globe. But more particularly it refers to arts and sciences, which in their nature, aim and results, are world-embracing. Engineers and Electricians do not represent nationalities or boundary lines. They represent that which is universal—Science—which is neither English or American—of which there can be no local appropriation, and concerning which there should be no territorial jealousy. Literatures may be owned. They speak different languages and breathe a confined atmosphere. Science cannot be claimed. She speaks one tongue for all peoples, and is the same in all climates and airs. Sir, our English and American flags are not broad enough, in their united folds, to cover the radical sentiment which animates the hearts of our own, and, I trust, the hearts of the English nation, in view of the recent union of the hemispheres. It is not the connection of two countries, that have never been really separated—for the umbilical cord was never cut, it has hung to us and kept us together, and it is this very day stronger than the Cable—No, Mr. Mayor, it is not the commercial and political pacification for all time of mother England and her great American daughter that now moves our chief joy and exultation. The Cable, more valuable as a symbol even than as a fact and an instrument, stands for a sentiment broader than our international territories, deeper than our dividing ocean—for an inter-hemispherical sentiment—the wholeness of a fractured globe, the unity of a scattered race, the sphering of a wandering humanity, the completion of the broken circle of *universal* hopes and sympathies and destinies.

Never before was anything purely human done in the history of the world and the race which stood for *One-ness* as the successful laying of the Atlantic Cable does! It is the beginning, on a scientific plane, of a reconciliation among things scattered and opposed, which is full of prophesies of peace and harmony through the whole internal and external world. Using reverently a word commonly mispronounced—wrested from its etymological significance, and so robbed of the pregnancy that belongs to it, but which one event like this throws a providential illumination on—I may say that the great principle of God's government and Christ's religion, At-one-ment, was never so majestically illustrated and employed outside of the very Gospel itself. Excepting ever a precious significance for the imagination, the uniting of the hemispheres of geography is the smallest meaning of this triumph. We have hitherto lived in a hemisphere, and we now live on a globe—live not *by halves*, but as a whole—not as scattered members, but as the connected limbs of one organic body, the great common humanity—and thus we are able to compass for the first time, as a popular thought, the conception of one life. one history, one interest for the race, one all-embracing, all inter-linking Providence—the establishment of a common, a right, a happy understanding among all nations, tribes, climes and interests, until the world of humanity shall move as the globe itself moves, all together; and the race be, as the individual is, vitalized by a spinal cord animating all its limbs, harmonizing all its movements, and giving coherency, energy and unity to its whole being.

A union like this reaches and includes many other now divided things besides the Old and the New World. The parted continents are not the only hemispheres that require to be pacified and united. Soul and body, mind and matter, conscience and inclination, work and play, business and leisure, duty and beauty, patriotism and philanthropy, interest and obligation, things secular and things sacred, time and eternity, earth and heaven, Man and God; these are the great intellectual, moral, and social hemispheres, which oceans, more stormy than the Atlantic Ocean, of immature thought, crude inexperience and coarse superstition have kept in seemingly hopeless isolation and antagonism. But the great process of bringing them together has begun, and this telegraphic triumph is perhaps the first stroke of science that fitly answers from the human side the mighty blow for man's reconciliation with his Maker, given 1800 years ago, from the superhuman side; for, like the Good News from heaven, it has aroused a triumphant hope that no oceans, whether material, social, or moral, can permanently resist the necessity for the tendencies to the providential plan of a Universal Harmony and Unity among all the interests of Humanity.

I know the misgivings connected with these commercial

and mechanical victories in the minds of many men of serious thought. I understand the part which "Materialism" plays in the vocabulary of ethical and religious discussion. As a nation, and by our own guides, we are exhorted to "put our hands upon our mouths, and our mouths in the dust," and to feel a special humiliation in view of the engrossment of our people in *material* pursuits! God knows the world has always needed cautions, and is never humble enough in view of its manifold short-comings. But I deny, before God, any special amenableness of our people to this charge, or any wisdom in the special form of this censure. Where is the *mind* of this country to be seen if not in its material results?—and what is so conspicuous in its material results as its *mind?* Is matter, turned to the service of the affections, the education, the culture of society, to be stigmatized and warned off the premises of the good? Are the great triumphs of art, industry, commerce, which are freeing, pacifying, blessing and elevating Humanity, to be denounced as material, in any sense of antagonism to morality, piety and the spiritual destiny of man? Why! what is this very victory of navigation, engineering and electric science, but a material victory—yet in what does it differ from or fall below an intellectual, a moral, a spiritual victory? Science, commerce, art, properly interpreted and providentially guided, have the same object, unconsciously or circuitously pursued, which religion and morality more directly propose. Commerce Christianizes by civilizing, whilst Religion civilizes by Christianizing; and both together work out the Creator's purposes and plans.

We are no longer, therefore, to talk as if the *matter* of this globe were opposed to its *spiritual* progress; it is to be made the instrument of that progress. We are no longer to concede that the business of the world is necessarily at war with the morality of the world, or to allow that a lasting internecine strife can go on among the various interests of man or the race. Our faculties, tastes, capacities, are a part and parcel of a nature thoroughly *one* in its constitution—our nature, position, material circumstances, part and parcel of a plan thoroughly *one* in its Divine conception and conduct. The Church, the State, the Home, the Workshop, the Ship, the Store, are helpers of each other—each sacred in its rightful place—each capable of a consecration such as has been displayed by the men engaged in this telegraphic work. They have felt—I see it in their eyes—that they were working, not for themselves, but for us and for God; and they laid the first offering of their hearts, when their enterprise succeeded, on the altar where all things first belong.

It is the sentiment of a Unity of which God is the centre, and about which, and in which, all things are circled—a unity now to be realized, as new successes remove, faster and faster, old causes of separation and dispersion — a unity which, commended to us in a signal and wonderful manner, by the recent telegraphic victory over space, ocean, disappointment, previous failure and present despair, has struck a thrill of peculiar joy and boundless confidence through the heart of the world. It is mingled joy and hope in the harmonious destinies of humanity that is now swelling the heart of the common people. No cheap feeling of national pride, no boasting exultation in the reflection that this work was chiefly accomplished by our own men, could produce this spontaneous enthusiasm. It springs from a source nobler, prouder, more glorious

It arises in the great suggestion that somehow "the beginning of the end" has come!—that, God helping, we have got to the end of the obstacles which threatened to prevent man's ever achieving a final triumph over those elements of discord and antagonism which lie in the path of human progress. We see now that a step has been taken which promises, nay, which assures, a continual progress, and whose end shall be the material and the moral redemption of the world.

This, Mr. Mayor and gentlemen, is the noble confidence which now swells the generous heart of this assembly and of the people of our country. Sir, there is no measure to the amount of moral and religious feeling awakened by this event in the remote hamlets and retired recesses of the land. Ten days ago I had a piece of the cable sent to me from New York, to exhibit at a village celebration in Walpole, N. H. An old man, of ninety years, sent down from the hills, and begged for a sight of the cable, saying "let me see the cable and die." He spoke like Simeon of old, and he spoke the sentiment of tens of thousands, who better realize their own immortality when they see the triumph of this material spark over time and space and ponder the mysterious and half supernatural continuity of the electric life. That spark is not material only; it is also Divine. God knows what difference there may be between the nature of that spark and of the spark that animates our own bodies; but if that can live through oceans, ours shall survive the grave! And in that confident feeling of life and immortality all the best hopes and animation for social and human improvement must begin and end.

I conclude with a sentiment which contains the pith of

my remarks, and which, however received, is offered in a deeply religious spirit:

"At-one-ment!—Science, Art, Commerce, Experience, Religion, have, in God's Providence, but one final cause—the at-one-ment of nature with man, of man with himself; of man with man, and of man with God." The Rev. gentleman resumed his seat amid warm applause.

The next toast was,

"THE STATE OF NEW YORK—May her history always illustrate her motto—EXCELSIOR."

To which his Excellency Gov. King responded as follows: The theme which the toast proposes for my reply is one full of deep interest and just pride to every citizen of this State, and especially to one who is to the manor born, whose youth and manhood have been passed within her broad limits, and who, during a life of many years, has watched the progress of her people and the increase of her power, until he beholds her now, in numbers, in wealth and in resources, the first among the prosperous States of the Confederacy—settled at the commencement of the sixteenth century, by the Dutch, who brought with them the love of Liberty and the spirit of commerce, soon to be succeeded by the English, who in their turn planted the same love of Liberty, the principles of the Common Law, and laid deeper on this favored spot, so well and so timely chosen, the foundation of a prosperous commerce. But it was left for the descendants of both races—their independence of the mother country being established, and the Constitution of the United States adopted—to understand and estimate the unrivaled position of this noble State for commerce, agriculture and the arts of peace. From that moment may be said to date her start in this contest for extended commerce; and at a later day, in her great works of internal improvement. With broad lakes and deep water on her western, northern and eastern borders, and the wide Atlantic pouring its tides around her commercial city on the south; with a fertile soil, an educated, numerous and prosperous people it may be truly said that the lot of her citizens is cast in a favored spot! This great advance, this rapid rise to power and greatness is due, next to her fortunate position, to the enterprising spirit of her people, to her fidelity to the laws and Constitution of her own and the General Government, and to the spirit of liberty which animates and ennobles her people and their undertakings. Steadily and proudly she presses forward in her march to prosperity, nobly seconded by the enterprise of her merchants, the skill of her mechanics, the thrifty labor of her husbandmen, and the wide-spread intelligence of her whole people. In peace and in war she has ever been true and loyal to all her duties and obligations, and names of renown and of honor are and have been among her public men. May no calamity, no intestine feuds, ever disturb or put in peril the happiness, prosperity and repose of such a community. But rather may her course be onward, ever generous, ever successful and independent. I would say a word in honor and in respect of this great city of the West—full of activity and of all the eternals of power and greatness. Noble and loyal city—the seat of commerce, the throne of those who compass the earth in the pursuit of honorable gains, whose streets are thronged with a busy and industrious people, whose public charities are nobly endowed, whose public schools are open to all—

may you ever be equal to your fortune and your position, and may the Almighty hand which has so far favored you, continue to watch over and defend you! I will conclude with the following sentiment:

"The City of New York—The great haven and mart of the Western World. Unrivaled in position and distinguished for the enterprise of her merchants, the skill of her mechanics, and the intelligence of her citizens. She feels and asserts the quickening impulse which commerce gives to the principles of liberty, to the spread of knowledge, and to enterprises of great pith and moment."

The ninth regular toast was:

"OUR CITY OF MANHATTAN—Foremost of America, now placed side by side with the chief cities of Europe; while we strive to surpass, we will be friends as well as rivals."

To this Mr. Richard Busteed, Corporation Counsel, responded. He said:

*Mr. President and Gentlemen:*

The extravagance of joy befits the purpose of to-night. The first city of the American continent speaks her congratulations to her sister cities, upon the accomplishment of the mightiest wonder of all time! The Submarine Telegraph—a fact of actual science—rests undisturbed upon its ocean plateau. A greater agency than steam is laid under contribution to the wants and will of man; a new means of increasing happiness and knowledge is furnished to the world. The subtle fluid of the gods goes upon the errands of men, and time and distance are figments of the past. The dream of the poet is realized. The mandate of Oberon is obeyed, and ere the leviathan can swim a league, Agamemnon and Niagara "put a girdle round about the earth."

We celebrate to-night the carnival of all religions and of all nations—the carnival of mind. "Our City of Manhattan, foremost of America, now placed side by side with the chief cities of Europe," spreads her municipal board in honor of the men and of the exploit.

It is meet, sir, that the great city which is our host, should thus commemorate the occasion. New York should not be silent—she could not be. Her people feel the electric thrill and are now shouting for joy. It should be so. No city in the world owes more to science and the arts. Sitting, a proud queen, upon her island throne, at the confluence of two rivers, her feet bathed by the ocean and her brow fanned by its breezes, she has become, and is, our nation's glory and pride. Commencing her municipal existence in 1652, with a population of seven hundred, she looks down to-night, at the expiration of but two centuries, and while yet in the infancy of her career, upon nearly a million of inhabitants.

Her commerce visits every sea, and the flags of all nations are unfurled in her harbor. Her wharves and piers, stretching out into the deep waters like the arms of a mighty giant, encircle in safe anchorage the shipping of the world. "Placed side by side with the chief cities of Europe," she emulates the zeal of all, and is outdone in enterprise and liberality by none. Whenever and wherever humanity calls her answering voice is heard; and whether it be to search in polar zones, mid regions of eternal ice, for a lost Franklin, or to supply a famishing people with food, her Grinnells and her Macedonians promptly respond to the demand. "While she rivals, she befriends her compeers."

Her noble charities, her public institutions, her hospitals, her homes for the outcast and the stranger within her gates, her Churches, her Universities of learning, her parks, her free-schools and her newspapers, the liberal and judicious provision which she provides for recreation and instruction, signalize her people as the true type of American Nationality. The city of New York is a condensation of the Republic. Liberal and enlightened in her government, tendencies and views, she draws to herself all that is desirable of restless energy and utilitarian activity. The mart for all the wares that brain or muscle can proffer or produce, she exacts thought, effort, labor, skill, *and she rewards them.* The Cosmopolitan city of Christendom—a social and moral Mosaic, whereof the characteristics of every clime, blended in exquisite harmony, form part. The ardor and fidelity of the Irish—the sturdy steadfastness of the English—the gravity and caution of the Scot—the vivacious and scientific genius of the French—the profound intellect and untiring industry of the German—the unpretending but effective labor of the Dutch—and the artistic capacity of the Italian: all these mingle with the genial fervor of the South and the indomitable perseverance of the North. In her streets the rich and poor meet and mix without invidious distinction. Honest poverty fears no offense, and receives none. Her high places are within the reach of all who have character, capacity and courage.

Without the test of birth or creed, she fosters all who seek her protecting ægis. Intolerance and fanaticism may momentarily disturb her quiet, but a healthy public sentiment soon re-establishes her tranquillity. Glorious city! "Foremost of America!" thy sons do thee homage! "Peace be within thy borders, and prosperity within thy gates." *Esto perpetua.*

The *City of New York* has especial interest in the subject of our present rejoicing. It was in the city of New York, twenty-three years ago, that Samuel F. B. Morse constructed the apparatus which demonstrated the practicability of the telegraph. In one of the rooms of the New York University this votary of Science incarnated the grand idea, and the first and fitting syllable of recorded lightning was—"Eureka." It was in the city of New York that the project of an Atlantic Telegraph was conceived. At the house of Cyrus W. Field the plan was born, nurtured and matured. New York was the first to feel the quickening impulses of this new nerve, and tingles now from centre to extremity.

Depend upon it, sir, our city will accomplish the prophecy of the toast. That prophecy is an inspiration. Brought nearer to the cities of Europe by this new agency, she *will* maintain an honorable friendship in the midst of an active rivalry. She will fulfill her destiny in a spirit of magnanimous pride, and with a just regard to the rights of competitors. Already the fifth city of the world, she must yet be the first. It is written in the book of fate. Thus in peace acquiring her prosperity and power, she will furnish encouragement and example for others, and sanctify her own greatness to the universal good.

I verily believe, sir, that the success of the Atlantic Telegraph will mark a distinct epoch in the history of the race. We have had the dark and the iron ages; the historian will signalize this as the telegraphic age. The consequences which are involved to man in this latest power are as yet but dimly developed. This magic network of material mechanism belongs to the future, and vibrates to *its* touch. It is inconceivably grand, and will find its highest triumph in this:—men will learn to war no more; pruning-hooks will take the place of spears, and swords be converted into ploughshares. In this millenium New York must share, and when the era which she inaugurates to-night shall be among the things that were—when a far-distant posterity shall occupy our places, and unborn generations enjoy the advantages of which this hour is the harbinger, how sublime will be the spectacle of nations sitting down together as do their representatives to-night, awarding the highest honors to those who peacefully increase intellectual victories.

When such an era comes to bless the world, posterity will find in the records and achievements of this hour something for pride and glory. Then shall people of every nation, and kindred, and color, and tongue, observe an inviolable amnesty; all hearts shall beat responsive to one pulse; *electricity will be merchantable, and insulated wire the world's amanuensis.*

Mr. Busteed concluded amid reiterated applause.

The tenth regular toast was then given:

"THE NEW YORK, NEWFOUNDLAND AND LONDON TELEGRAPH COMPANY—which commenced and planned, and the

ATLANTIC TELEGRAPH COMPANY, which completed, the work of linking two continents together beneath the sea: they have done more for the civilization and peace of the world than any other companies which ever existed."

This toast was responded to by the Hon. Augustus Schell, Collector of the Port of New York. He said:

I feel honored, Mr. President and gentlemen, by your call to respond to the toast you so warmly and justly applaud. I shall not detain you by dwelling upon the great features of the stupendous result which we are assembled this day to celebrate, and which commands the admiration of rejoicing nations, nor by tracing step by step the labors and discoveries of those earnest and patient men of thought—Franklin, Oersted, Ampere, Arago, Farraday, Jackson, Henry, and our distinguished and illustrious Morse—to whose developments telegraphic science and this gigantic climax are so largely indebted. These, though connected with my theme, will be topics which others of more congenial pursuits and greater leisure will do eloquent justice to. But I will bring to your consideration a few words upon the immediate subject of this well deserved tribute to the New York, Newfoundland and London Telegraph Company and the Atlantic Company, one of which commenced and planned, and the other completed, the link which binds two continents together beneath the sea. Early in the spring of the year 1854, five gentlemen of New York met at the residence of one of their number—a well known and liberal-hearted citizen—and laid the foundation of a company, which after four years of doubt—of almost despairing effort—of seeming inevitable failure—has astonished the world with an achievement so gigantic in its design, so momentous in its consequences, that the mind almost loses its equilibrium in the contemplation of it. These five gentlemen were Cyrus W. Field, Moses Taylor, Peter Cooper, Marshall O. Roberts and Chandler White—the residence, that of Mr. Field. Here, with much confidence, but more hope, they made the first subscription and assumed the first responsibility of the great enterprise. The comparatively modest sum of ten thousand dollars a piece was thought would be sufficient to start, if not prosecute the work. How inadequate this starting sum, when, after repeated failures and darkening prospects, the outlay of each of the survivors of this brave band reached nearly a quarter of a million of dollars! Early in the spring of 1854 Cyrus W. Field, Chandler White and David Dudley Field were sent by this incipient company to Newfoundland to procure from the Governor and Legislature of that province, then in session, a charter with certain privileges and powers. The Governor promptly responded to the voice of the commission, called a meeting of the Executive Council, and transmitted a message to the Legislature recommending the incorporation of the Company, a provisional guarantee on the interest on the bonds of the Company, and a grant of fifty square miles of land in aid of the Company. At once the Legislature gave its sanction to the recommendation of the Governor and Council. In three days the commission returned to New York with the charter, guarantee, grants of land and privileges. Whether owing to the perseverence of Mr. Field and his associates, or guided by an enlightened and rare forecast, the authorities of Newfoundland have, from the outset, evinced, and continue to evince, the highest appreciation of the purposes and objects of the Company; and to their alacrity of action and munificent donation do we owe much of the early and latter success of the great project. Two years after this occurrence the Company, desirous of securing the co-operation of English capitalists, dispatched Mr. Cyrus W. Field to London. England, not less than Newfoundland, yielded at once to the grandeur of the idea, the cogency of his reasoning, and the vigor of his efforts; and the balance of the entire capital was at once subscribed, and through his agency another and auxiliary Company was chartered by an act of Parliament. Its shareholders embrace many of the leading scientific and political minds and greatest capitalists of both countries. The two companies in purpose and object are, I am informed, a unit, co-operating in the expenditures and efforts, and sharing mutually the honors and rewards of the glorious consummation. Besides the subsidies I have mentioned, Newfoundland has made an additional grant of fifty square miles of land to the Company when the Cable shall be laid from Ireland to Newfoundland. The large subsidies, also, of the English Government and our own, and the active interest taken by both in the success of the enterprise, are well known to you all. Under the favorable auspices the Company enters upon its career of usefulness, what will be its destiny—what its progress or profits, I shall not venture to calculate or predict. Let us at least wish for them a successful management and enduring prosperity. Among the many names eminent in scientific experiment and discoveries, of whose labor this is the crowning result, it will be seen that our country can point with just pride to an honorable proportion of our countrymen; and whatever may be its effect on the great business commerce of the world, no one can doubt but that our own city will participate as largely as any other in its substantial benefits. As nothing comes by chance, it may be said that this marvel of our times is the result of particular agencies which have existed for a long period of time. The submarine plateau—the milky-way of the bottom of the sea—throughout of a right and safe depth and width, forming a bed of minute shells for the security of the submerged Cable, and a substance discovered in a distant land by which alone insulation could be secured, besides many concurring corollaries, are of marked significance, and teach a lesson in our rejoicing, that while we employ human agencies, we should not forget the exalted source from which sprang, as if by a direct emanation, this resplendent, instantaneous track of mind from continent to continent, from nation to nation, throughout and around the habitable globe, lending wings not only, but giving ubiquity to thought and events, causing everywhere the profoundest sensation of wonder and delight. Mr. President and gentlemen, I congratulate you that we live in an age of so much grandeur of discovery and accomplishment, of such exalted consummation, so full of promise to the peace of nations, of enlightened progress and of coming illumination, and that we are permitted to rejoice with far-off nations as if they and we were at the same festive board; and that we can, by a flash across the wires, report even to them the justly complimentary toast to these Companies at the moment we respond to it here.

To the eleventh regular toast—

"The Arts of Peace—Now crowned with immortal lustre and for once at least covered with greater renown than ever were the Arts of War,"

The HON. DANIEL E. SICKLES, M. C., responded :

He concluded an address of much interest by reading the following "Song of the Flags," which was composed by the Rt. Rev. Bishop Doane, of New Jersey, and forwarded by Telegraph to Rev. Dr. Ogilby, of Trinity Church, with a request to have it read at the Dinner :—

THE WEDDED FLAGS—A SONG OF THE CABLE.

(The English and American flags, displayed together from the spire of Trinity Church, were blown across each other in a mutual embrace.)

Hang out that glorious old red cross—
Hang out the Stripes and Stars !
They faced each other fearlessly
In two historic wars.

But now the Ocean's circlet binds
The bridegroom and the bride,—
Old England and Young America !
Display them side by side.

High up from Trinity's old spire
We'll fling the banners out ;—
Hear how the world-wide welkin rings
With that exciting shout !

Forever wave those wedded flags
As proudly now they wave !
God, for the lands his love has blessed
The beauteous and the brave.

But see ! The dallying wind the Stars
About the Cross has blown ;
And see, again, the Cross around
The Stars its folds has thrown.

Was ever Sign so beautiful
Hung from the Heavens abroad ?
Old England and Young America
For Freedom and for God !

Mr. JAMES BROOKS responded to the 12th regular toast, which was:

"THE PRESS—To which the Telegraph is both minister and instrument; may its usefulness be always equal to its powers."

He said:

I have accepted, Mr. President, the agreeable duty of responding for the Press, because my profession of journalism is under the greatest obligations to the magicians who have realized not only the lover's but the journalist's prayer, in the annihilation of time and space; nay, who have done more in giving us in this Western World, the news of the morning long before, in the translated words of Homer, Aurora

"Has sprinkled with rosy light (our) dewy lawn."

As journalists, now we can sit in our attics, and there leaps in upon us, through the coral groves of the oceans, the yet live news then buzzing through the alleys of Threadneedle street, London, or surging against the dome of St. Paul's.

Magicians I have called those men; and are they not magicians, who can bring to us at noon-day, on the banks of the Hudson, the doings of the antipodes, of the Brahmin, the Hindoo, the Mohammedan, on the Indus, on the Ganges, and who can so closely connect the antipodes, as to make us fancy we really see the rays of the morning sun that purple the peaks of the Himmalay, gilding at the same moment the evening clouds that hover over the

Alleghanies? The Genius of news, the newsmen, the printer, the publisher, no wonder, are inspired!—and all the more they rejoice, because it was Franklin, the printer, who, in the glowing words of his eulogist,

"Eripuit fulmen cœlo, sceptrumque tyrannis,"—

Franklin, the printer, who caught the lightning in the clouds, sported with it on the kite, and made it run, dancing over his lightning-rod, into the earth. And, because it was Morse, the printer, who took the lightning into his infant school, taught it there its letters, then how to speak and how to spell! And Field, the paper-maker, the paper-seller, who wound his cable around the blustering arms of Old Neptune, and from the rogue wrenched the Trident of the seas!

Three great events, Mr. President, link the Old World and the New. In 1452 the world was only Europe, Asia and Africa; and what of that world was not on the shores of the Mediterranean, was about all in Castile and Arragon. Then, Christopher Columbus left the little port of Palos in his caraval, a coasting craft of some ninety tons, and, landing on San Salvador under the banner of the cross, handed over all America to Ferdinand and Isabella. The wind—the latten sail—did all this. In 1838 the little British steamer Sirius, Capt. Roberts, as bold as Columbus, ventured from the highlands of Britain, and in spite of the wind, anchored off our Battery, before a multitude so astounded as if in these dog-days the bright dog-star Sirius had dropped from its constellation of Canis Major upon that Battery. Steam then conquered wind and ocean wave, and Old Æolus was hushed up. But time had only been shortened, and there yet existed space! Then started the conception of Field; then Maury and Berryman mapped out the bottom of the sea; then the genius of Morse; then the generous capital of New York and London's princely merchants; then the glorious story of the Agamemnon, the Valorous, the Niagara, the Gorgon—not

"Gorgon and chimera dire,"

—and time and space were annihilated!

Now, who can see or foresee the result of all this? Cotton, which boasts of being "king," feels in this extended electricity new pledges of sovereignty. Rice is exultant. Tobacco even is not without joy. Gold, ever glittering, glitters yet more and yet wider in the sparkles of the wire. But I see, or think I see, it is the Printer who is to win the day, and through him, the universal RESPUBLICAS. The Electric Telegraph is but another way of printing. Its business is Letters. It lives on words, thoughts, ideas; and is a pure spirit—in nothing material. Through it and by it, our American thoughts, American principles, American precedents are to flash daily into the dusky rooms of St. James, the glittering halls of the Tuilleries, the majestic grandeur of the Escurial, the Austrian Schonbrunn, and the Roman Vatican.

The minarets of the mosques of Constantinople, and the towers of the Kremlin are to be lit up by them. No Horse Guards, no Corps Municipal, no fiery Sun nor wild Cossack, no crack of Minnie rifle, no crash of shell or bomb can fence off this spirit imponderable, invincible. The Queen, the Czar, the Emperor, must use it as they use the air all breathe. Hence then comes a mighty conflict of antagonistic principles—an equalization of thought, the fraternity of mankind. Something is to topple over—but whose that *something* is, no Cassandra at a dinner-table like this can predict.

But sons of the new world, as we are, the pleasing reflection of such a re-union as this—whether we be Anglo-Saxon, German, French or Spaniard—is the living link we now have with our fatherland. The German, the world over, it is said, whether on the Volga or the Mississippi, chants the song of Wolfgang Muller.

> Mein Herz ist am Rhein, en heimischen Land,
> Mein Herz ist am Rhein, wo dei Wiege mir stand.
>
> (On the Rhine is my heart, where affection holds sway,
> On the Rhine is my heart, where encradled I lay.)
> Wo ich bin, wo ich gehe, mein Herz ist am Rhein.

But oh, with how much more enthusiasm may we Americans, Englishmen, Irishmen, Welshmen, Scotchmen, parted in British Colonies as wide apart as the poles, lay our hands and our hearts too upon the Ocean Cable, and feeling its pulsations, turn our eyes homeward to our vaterland, singing forth "Home, sweet home, there is no place like home." We, who first lisp together in the words of Shakespeare and of Milton—we, whose mighty fathers lie buried together in the consecrated aisles of Westminster Abbey—with how much more enthusiasm, I say, may we, now the globe encircling in words and wire, as the earth rolls on its axis—sing together, the live long day, in one continued melody, the spirit-stirring words:

> "God save the Queen."
> "Hail Columbia."

The thirteenth and last regular toast, was—

"WOMAN—At whose feet we lay all our triumphs; to her we owe the happiness of life and the consolations of home. God bless her."

Hon. John E. Ward, of Savannah, Ga. (President of the Cincinnati Convention that nominated Buchanan), replied. He said he did not know how he should respond to such a sentiment, so far from the consolations of which the toast spoke—Home. But they might be forgotten in the happiness which the Cable demonstrations here had given those who had come from afar to witness or participate in them. Great as were the binding ties of the Cable, there was a greater tie between men of the two hemispheres—Woman. In proper terms he alluded to the virtues of the Queen of England, and declared that every American citizen respected the smile which played around her lips. Woman's consolation, he concluded, would ever draw man to home, or console him in absence from it.

The following volunteer toasts were then offered:

By Mr. CYRUS W. FIELD: "The union of England and the United States in the enterprise of uniting the two hemispheres—illustrated by the exertions in its behalf of Wilson G. Hunt, an American, and Edward M. Archibald, an Englishman."

Loud calls were here made for those gentlemen, but they were both absent.

"To Wm. E. Everett, the Engineer who laid the Cable—by whose inventive genius the triumph was achieved."

To this the Hon. John Cochrane responded:

It is not very often, sir, that praise is bestowed upon what are generally esteemed the secondary instruments of any great achievement. The glory of success nearly always envelopes the name through which it is known, and the subordinates, though indispensable, are sometimes forgotten. I cannot, however, ascribe these remarks to the fortune of those engaged, in however humble capacity, in the laying of the Atlantic Cable. An enthusiastic public has lavished deserved honors upon them all. They all have received bountifully of the popular plaudits; and, certainly, if due appreciation of unremitting care and superior skill forms any part of their reward, it has been granted with an alacrity excelled only by the modesty with which it has been received. The toast, sir, just offered is designed to honor one of the chief actors in the expedition which we celebrate. I suppose that the nature of his services is generally understood. Yet, it were not amiss to recur to them specifically at this time. I believe that the impression prevails that had an efficient machinery regu lated the Cable in its descent to its bed during the effort of 1857, it would not have been unsuccessful. At all events, it is indisputable that in the interval between the first and second attempt, constantly invention was charged with the discovery of the means for accommodating the strain upon the Cable to its strength. This discovery is due to the genius of William E. Everett. We are informed that the search had been unsuccessful, and that the most sanguine friends of the enterprise began to despair, when the invention of Mr. Everett was first explained to their incredulous understandings. His practical ability disclosed to the unpracticed the merits of his machinery, and repeated experiments at length attested its competency. Thence the laying of the Cable was but a question of time. With the means at hand of depositing it from on shipboard in profoundest depths, under all strains occasioned by every weather, whether an accident should intervene was but the conjecture of a possible delay, while ultimate success became only the product of labor and perseverance. Behold the fleet, then, upon its voyage. The splice has been made in mid-ocean. From two vessels, moving steadily from each other, proceeds the uniform run of the Cable. Whether on the billow or in the calm, its speed is controlled and its integrity preserved by the machine created by the master's genius, with power to deliver to the order of Old Neptune in installments, as required, the gutta percha, and the copper, and the steel, with which to pave his rugged ocean floor; and so they take their eventful way—that one to Valentia Bay, and this one to Trinity Bay—and when each goal has been reached, and the last coil has been paid out, why there it stands, in impressive silence, indicating to an enraptured world, by the work it alone has done, the great truth that, as often as employed, it can do it again and again, through every wave, in every sea. I claim, then, great honor for the inventor of this successful Cable layer—of this great Cable king—honor for Wm. E. Everett. Permit me, Mr. President, while upon this theme, and in further illustration of it, to refer, however briefly, to the essential offices of mechanical genius in the accomplishment of every enterprise. The scheme of speculation, and the vision of philosophy, however practical or real, can be attained only through the instrumentality of the mechanician. The most simple scientific principle is nugatory save when subjected to the harness of mechanics. Thus applied, it constructed pyramids in Egypt, it destroyed navies in Syracuse; and thus applied, it turns the spindle, it drives the hammer, and it moves the wheels that bear along the massive superstructure of our modern social fabric. Your mechanic is your true hero. He is in your workshops with his patents, in your fields with his machines. He is to be found in your mines, teaching men how to prepare the metals for use; he is to be found on your hills, directing man how to remove them from his path. He rides upon the waves and chains them fast; he dives to the bottom of the sea, and, at a thought, subjects it to human dominion. He is the motive power of the world. Though science may demonstrate what can be done, it can be done only by mechanics. Though philosophy may teach that nothing is impossible, mechanics alone can teach *how* nothing is impossible. It is a beautiful and instructive allegory that—familiar to our school days—where the Cretan nymph possesses the Grecian hero with a clue to the intricacies of the labyrinth through which he removed the symbolized obstacles to a union of Attica with Crete. May we not apotheosize the hero who has laid the clue possessed of which all the world again is kin? Truly, sir, we do not err when magnifying the far reach of his comprehensive mind, who of his philosophy conceived, and of his sublime faith confided in, the theory of a western world; and yet what had been the fruits of that grand thought—what the destiny of that grand old mariner in his pathless way on the deep, had not the compass taught him how to "steer securely and discover far?" And where, let us speculate, had still been the abode of the lightning—of this subtle spark we are now dispatching all over the earth and all under the sea—had not the mechanical genius of Franklin come to the aid of his philosophy, and instructed him how to attract the volatile essence, and to lay it under bonds to man? Sir, many another instance is there which approves the mechanician to be the true benefactor of his species. Honor, therefore—again I say honor to the man whose genius taught how the Cable could be laid, and it was laid. Sir, there are, to my mind, three eras in the history of word-traveling. The first introduced an electric current which bore instantaneously away beyond recorded distance, over the mountains and beneath the seas, human thoughts to other human thoughts. The second saw men moved to gracious accord, capital collected and bestowed, and labor and perseverance applied to the accomplishment of the noble enterprises of the Telegraph. The third witnessed the humble and careful approach of the mechanical arts to the assistance of the greater science, and behold the consummation. There they stand—Morse the inventor, Field the executive, Everett the mechanician; and so will they be held in enduring remembrance.

By Mr. Joseph Hoxie: "The late naval engagement between her Majesty's war steamer Agamemnon and the United States frigate Niagara—the only one on record in which both were victorious."

The following toast by John D. Jones, Esq., was then announced:

"The United States Coast Survey and Lieutenant Berryman—The nation will not soon forget their disinterested efforts in favor of the enterprise, the success of which we now celebrate."

Prof. Bache responded, and defended the Coast Survey from the anonymous charge of opposing the great enterprise, because its (the Coast Survey's) chief men had predicted failure, except at the time when the enterprise proved successful. He said:

Gentlemen—I wish I could say gentlemen and ladies —I read in a paper, published perhaps in Albany;—yes, it was the organ of the triumvirate who govern the Trustees of the Dudley Observatory, (Trustees, *lucus a non lucendo*,)—that the Coast Survey was opposed to the Atlantic Telegraph, and you, emphatically its friends, toast the Coast Survey. There is certainly a mistake somewhere. Let Mr. Field, Mr. Cooper, and Mr. Hunt decide where it *lies!* It is not given to all to die for a cause; many, many live for it, and more may serve it. We have the highest authority for knowing that one or two talents, rightly used, may meet with appreciation and even with reward. Consistently, from the beginning to the end, has the Coast Survey been ready, by all lawful means, to serve this great enterprise. Witness the last year's labors of Berryman. Witness other unpretending efforts well known to Matthew and to Cyrus Field. None have rejoiced with a deeper sympathy than the officers of the Coast Survey in the glorious success which has crowned the undertaking. For myself, I can truly say that no heart glowed more warmly and was more filled (beyond these of the hallowed family circle) by the glories of the civic triumph of yesterday, than my own. True, the triumvirate say that heart is bad, I am an unscrupulous man, a tyrant. What say the officers of the Coast Survey to this? Is there not some mistake, my friends? Has a tyrant *such* friends?

In our country, in the last century, a party of scientific men and of friends of science, inoculated by the warm zeal of Franklin in relation to electricity, assembled on the banks of the Schuylkill, near Philadelphia, to hold an electrical feast. Our ancestors were quaint in some of their sayings and doings. Part of the day's programme—(they didn't call it so)—was to pass an electric charge through the river, and by the shock to procure the dish that was to grace the head of the table. It was on this occasion the laughing philosopher said that the people might pronounce that there was a force at both ends of the electric chain—but no they did not—they laughed, and they made a myth, a demi-god of him—seeing in the experiment a full proof of the great principle of transmission of the electric current through an expanded conductor. Upon what a magnificent scale has this principle been applied in this century! Instead of a few simple philosophers meeting in sport or mirth, the representatives of the two great nations exchange greetings of solemn import, of peace and good-will. Instead of the Schuylkill is the wide Atlantic. Our Field is the World, and the World's. And yet the popular voice, speaking through the mottoes of yesterday, acknowledge the giant of science (Franklin,) while giving every honor to Field, and Morse, and Hudson, and Preedy, and Dayman, Whitehouse, and Bright, and Brett, and Everett, and Woodhouse, and many more, for there is glory enough for all. Why did not this voice recognize the claims of that son of New York who, when Barlow found the telegraph impossible, determined the laws which rendered it *eminently practicable*—Joseph Henry? To every one his meed.

Gentlemen, you hardly thought that Franklin had any direct agency in laying this Cable. But I have it from a man, who rivals the great barber of the Arabian Nights as a talker, and excels him as a philosopher, that an ingenious customer of his assures him that Franklin has devoted himself, in the Spirit world, in an especial way, to this undertaking, and that, being recently aided by one whose loss in this upper world we still deeply deplore, he has brought this enterprise to a successful issue.

Gentlemen, the survey which you have just honored now presents a continuous triangulation from Machias to the northern boundary of South Carolina, and thence, with a few links only wanting, to the St. Johns. If we could have the prosperous times which preceded 1857, we should show you in twelve years a land and water survey, completed, of the whole Atlantic and Gulf coast, with considerable progress towards completion on the Pacific. We have rendered it certain that the entrance to your port has not deteriorated, but has rather improved, within the last twenty years, and have furnished the materials by which your own citizens have secured your noble harbor from suicidal encroachment. The nation sees and hails the future of this great port, when, connected fully with the Pacific by railroad and telegraph, as it is now with Europe by the Cable, it shall become the commercial centre of the world! Emphatically, with the Telegraph, there is no East, no West—only a centre. Time itself shall be no more!

Amid the thunders of Sinai the still small voice could be heard, and amid the roar and din of the world's strife and contention, comes the voice of peace from mother and daughter, and the quick, sympathetic response, the electric kiss of America and Great Britain.

I used to *think* triangulation ("nothing like leather") was the great connector of capes, and headlands, and continents, but now I know it is the Cable, and we of the Coast Survey, under the enlightened head of the Treasury Department, have been ready now these two years and more, standing in eager expectation over our transits and chronographs, "double shotted, pricked and primed, port fires ready, match lighted," waiting only the word from Field, and Cooper, and Eddy (illustrious name of a great and good, a simple and single minded man, gone to his rest without seeing the glories of yesterday), to fire! and our range is from America to Europe. When the winter's campaign in Washington, which our good, kind friends of the Dudley Observatory so blandly promise us, shall come off, we will "beat to quarters," rally the merchants and navigators, the ship-owners and shipmasters of New York and of the country, and make a great defense under the Cyrus of the Nineteenth century (may he find a Xenophon!) with the rock Peter ready for use in our ballista.

Mr. Peter Cooper then rose, and made the following remarks:

Gentlemen—It will be impossible for me to give form and expression to feelings that struggle within for utterance. The occasion which has called us together is one of such vast and immeasurable importance to the world that the minds of all are bewildered in the contemplation of its results. That an electric power has penetrated the world's heart, forcing pulsations of sympathy into every fibre of the body, inspiring the people of every region with a joyful hope of a brighter and better day for the world, is a fact which may well tax the strongest and most facile powers of expression to convey its full import. The day will come when knowledge shall take the place of ignorance—when science shall have developed the laws and methods of Deity—when Christianity shall spread over the world, having its foundation on God our Father, and the world of mankind be brethren. Then a universal charity will arise, growing out of a better knowledge of the power—the mighty power that the circumstances of birth, education, and country have exerted in the formation of all the differing characters and conditions of mankind. I will not, gentlemen, occupy your time with a history of the rise and progress of the company which I, in part, have the honor to represent. It is sufficient to know that the labors, expenses, and hazards encountered in the progress of the work have at last electrified the world with success—a success that will shed an enduring lustre on the enlightened Governments that have lent their powerful aid, and on all that noble band of officers, sailors, electricians and engineers, by whose united efforts the glorious prize has been secured for the world. May they ever be rewarded with

> "The soul's calm sunshine and the heartfelt joy
> That nothing earthly gives or can destroy."

Gentlemen, I will give you as a sentiment:

"The Atlantic Cable—The longest, strongest, and surest bond to keep the peace of nations."

Several other volunteer toasts were given, among them the following:

"FRANKLIN AND MORSE—The one was a Prometheus, the other a Cadmus to the lightning of Heaven; one brought down its forked javelin from the clouds, the other made it a pen and taught it the syllables of human speech and universal brotherhood."

By James Harper, Esq.:

"American Enterprise—Give it a fair Field, and it will be sure of success where Everett goes."

By General Hall:

"The Chimes of Old Trinity—They are the same old bells presented to Trinity Church when the diocese was first instituted, and for the first time in three-quarters of a century, they pealed forth the national anthem of Great Britain, "God save the Queen."

By Henry O'Reilly, Esq.:

"Lieut. Maury, of the American National Observatory—The indefatigable investigator of the winds and currents, and soundings of the ocean, which have made his name known and honored throughout the world; whose researches have occasioned the designation of that wonderful submarine 'Plateau,' which will forever bear his name in connection with that glorious success of the 'Atlantic Telegraph,' which the whole civilized world is now simultaneously celebrating."

Mr. Charles H. Haswell, President of the Board of Councilmen, being called upon for a toast, after a few appropriate remarks, gave

"The health of Professor S. F. B. Morse."

By Homer Franklin, Esq.:

"The Anglo Saxons of the East and West—the Joint Committee on the World's Commerce. They report progress 'on a string.' May their future 'reports' be equally creditable to England and America."

This was followed by

"Captain Otter, who lighted the fires and led the way to Trinity Bay. We do him honor."

"The New York Press and its attaché upon the various Trans-atlantic Telegraphic Expeditions. All honor to the 'Historian of the Expedition.'"

Mr. Clancey, President of the Board of Aldermen, being called upon, introduced Mr. Augustus J. H. Duganne, who recited the following ode, composed by himself for the occasion:

## HYMN

### For the Laying of the Atlantic Telegraph Cable.

BY AUGUSTUS J. H. DUGANNE.

Oh, Jehovah! oh, Elohim! be the glory all thine own!
For the stars in marvellous courses are but Voices from thy throne!
And the zones of mortal dwelling, and the oceans as they roll,
All obey Thee, all adore Thee, Master of the Immortal soul!

Thine the chart the Chaldean pondered, 'mid his orient skies unfurled,
Thee the tortured Galileo poised above his "moving world;"
Thee Copernicus, enraptured, magnified, with dying praise,
And the adoring Newton saw Thee—Ancient of Creation's Days.

Thine the Name—oh, Lord of Wisdom—Thine the Word of Life divine;
First, in mystic joy and trembling, matrized by the German Trine,
While the souls of mouldered ages, in their old Imperial dress,
Walked, in grand transfiguration, through the portals of the Press!

Lo! the sunbeam limns our features; Fire and Air we yoke to toil:
Yea, the lightning from the footstool we have chained in hurtless coil!
Thou, oh, God, o'er FRANKLIN bending, gave to him th' electric flame,
And with cloven tongues exultant, MORSE declared Thy Holy Name!

Scrolled beneath the sundered ocean—scored by lightning's awful pen—
"Glory unto God, the Highest! Peace on Earth, Good Will to Men!"
Land to land, in mingling currents, sways and thrills with loving fear,
"Where art Thou?" the Old World whispers, and the New World murmurs "Here!"

Here th' elastic Heart of Nations—here th' eternal core of Right:
Radiant from their burning centre flash the veins of Freedom's light!
Girt with all the world's great waters—circled far by all the lands—
Marked by sacred Line and Plummet—God our destiny commands.

Father! God! we faint—we falter! Lord of elemental powers—
Grant us that, with God-like wisdom, child-like humbleness be ours!
Thou hast made mankind vicegerent o'er the realms of mind supreme
Be our hearts Thine earthly altars—be Thy wondrous love our Theme

After which, the party broke up, all well pleased with both the material and the mental entertainment that had been afforded them.

Among the features of the evening was a telegraphic machine of the American Telegraph Company, which was placed in the corner of the room, just inside the door, and was kept in constant operation all night by Mr. J. K. Calvert, who transmitted and received several messages to and from Halifax for Mr. Field and others. The machine was connected with the telegraph wire on Broadway, in expectation of the receipt during the evening of a message from the Lord Mayor of London, but it did not arrive.

## THE LAST PYROTECHNIC DISPLAY.

Whilst the guests at the dinner table were indulging in the multifarious luxuries which the larder of the "Metropolitan" affords, the populace were feasting their eyes on a display of fireworks at the City Hall. This exhibition was prepared under the superintendence, and at the cost, of Mr. John W. Hatfield, and was witnessed by an immense throng which crowded the park, and every place adjoining, from which anything of the exhibition could be seen. The programme, as carried out, was as follows:—

1. The commencement was announced by the discharge of signal rockets, followed by balloon ascensions.

2. An eccentric piece, opening with a double wheel in various colors, changed to a pyramid of horizontal wheels, which, after assuming numerous changes, terminated with a beautiful representation of a weeping willow.

3. Flights of shells of various colors.

4. The Star of America, beginning with a vertical wheel in green and gold, changed to the Star of America in silver lancework, with crimson rosettes, and concluded with a double star in golden fire.

5. Batteries of colored candles.

6. A beautiful mechanical piece opened with three horizontal wheels; changed to three vertical globes, which, by their combined motion, represented the annual and diurnal movements of the earth, showing the various lines in scarlet, purple and green fires.

7. Display of bomb-shells.

8. Flight of rockets.

9. An appropriate motto piece, opening with a vertical wheel; changed to a motto suitable to the occasion.

10. Batteries of fire pumps.

11. Grand gallopade of serpents—an extensive gyration piece, the centre portion of the device representing four large serpents, each introducing four smallar ones, turning reversely, and assuming a variety of lively and fanciful changes.

12. Display of shells in gold and silver fires.

13. A most superb and extensive mechanical piece, opening with circumfused wheels, changing to an elegant scroll in silver lancework, surrounded by revolving suns, exhibiting an immense mass of moving fire, and showing at one view all the various colored fires at present known in the art of pyrotechny.

14. Immense flight of rockets of various colors.

15. The Kaleidoscope; commencing with a double triangular wheel in red, green, and mazarine blue fires; changing to the kaleidescope, which is formed by the combined revolutions of various complicated figures; concluding with a double revolving sun in Chinese fire, reported.

16. Volcanoes of colored fires.

17. Shells of stars, serpents, gold rain, &c.

18. A curious cycloidal wheel. Commencing with a large scroll wheel of every possible color, changing to the cycloid; cycloid formed of innumerable rings of every conceivable color, and representing a most intricate and pleasing device, formed by the mechanical movements of the various fires employed. Concluded with a geometrical figure in brilliant fire, marooned.

19. Motto piece, in brilliant colored lances.

20. Brilliant illumination of the entire front of the City Hall.

21. Grand finale. The laying of the Atlantic Cable. In this piece some new principles were attempted for the first time in the pyrotechnic art; the open portion represented the Niagara and Agamemnon in the centre, with the tenders Gorgon and Valorous ahead; on the extreme ends of the piece were two light houses connected by a line of rolling waters, on which the ships slowly moved towards their destination; on arriving at which the centre was suddenly transformed into a magnificent temple of science, in all the splendor of the dazzling colors, assisted by all the mechanical contrivances of which the art is capable. Canopied by an arc of stars which rested upon revolving columns, upon whose base were recorded the names of Franklin, Morse and Field, appeared a group of figures representing Science uniting Columbia and Britannia. Over these was an entablature with the motto: "The electric flash shall belt the earth." The crowning portion showed the American coat of arms, in which was entwined the Union Jack with the Stars and Stripes.

In order to do honor more fully to this great success of

science over time and space, it was found desirable to employ electricity, thus creating an actual as well as an imaginary current through the entire length of the piece.

The whole concluded with batteries of candles, flights of rockets and bomb shells, filling the air to a great distance with colored stars, gold rain, fiery meteors, serpents, &c.

And so, in the city of New York, were terminated the public rejoicings in honor of the successful submersion of the Transatlantic Cable.

---

On the 11th November, the following communication from his Honor the Mayor was received by the Board of Aldermen, and referred to the Joint Committee on Celebration:

MAYOR'S OFFICE,
NEW YORK, November 11, 1858.

*To the Honorable the Common Council*:

GENTLEMEN—I have the honor to transmit to your Honorable Body a copy of a communication received by me from Lord Napier, Her Britannic Majesty's Minister Plenipotentiary, at Washington, expressing to the members of the Common Council and myself the thanks of Her Majesty's Government for the part taken, at the celebration and festival in commemoration of the first establishment of electric communication between Great Britain and the United States, by the civic authorities of this city in regard to her Majesty's officers, as well as for the manifestation of friendly feelings on the part of our citizens towards Great Britain.

I need not add the pleasure I experience in submitting this communication of Lord Napier, manifesting as it does those friendly sentiments which so happily exist between our own country and the great nation which he so ably represents.

DANIEL F. TIEMANN,
*Mayor.*

(Copy.)

HER BRITANNIC MAJESTY'S LEGATION,
WASHINGTON, November 7, 1858.

SIR—It was my agreeable duty to submit to Her Majesty's Government some account of the festival, held at New York, in commemoration of the first establishment of electric communication between Great Britain and the United States, and to remark upon the cordial sentiments which were evinced on that occasion by the Municipality and the inhabitants of the city towards the Sovereign and the people of England.

I am now instructed to express to your Honor and to the members of the Common Council the thanks of Her Majesty's Government for the part taken at that celebration by the civic authorities in regard to Her Majesty's officers, as well as for the generous manifestations of friendly feelings on the part of the citizens towards Great Britain.

I have the honor to be, Sir,
Your most obt. humble servant,
NAPIER.

www.ingramcontent.com/pod-product-compliance
Lightning Source LLC
LaVergne TN
LVHW010226110826
845151LV00004B/1192
* 9 7 8 1 4 2 5 5 2 6 3 0 6 *